Machines that Think

Machines that Think

Everything you need to know about the coming age of artificial intelligence

NEW SCIENTIST

**New
Scientist**

First published in Great Britain by John Murray Learning in 2017
An imprint of John Murray Press
A division of Hodder & Stoughton Ltd,
An Hachette UK company

This paperback edition published in 2022

1

Copyright © *New Scientist* 2017, 2022

The right of *New Scientist* to be identified as the Author of the Work has been asserted by
them in accordance with the Copyright, Designs and Patents Act 1988.

All rights reserved. No part of this publication may be reproduced, stored in a retrieval
system, or transmitted, in any form or by any means without the prior written permission
of the publisher, nor be otherwise circulated in any form of binding or cover other than
that in which it is published and without a similar condition being imposed on the
subsequent purchaser.

A CIP catalogue record for this title is available from the British Library

B format ISBN 9781529381955
eBook ISBN 9781473629660

Typeset by KnowledgeWorks Global Ltd.

Printed and bound in Great Britain by Clays Ltd, Elcograf S.p.A.

John Murray Press policy is to use papers that are natural, renewable and recyclable
products and made from wood grown in sustainable forests. The logging and
manufacturing processes are expected to conform to the environmental regulations of
the country of origin.

John Murray Press
Carmelite House
50 Victoria Embankment
London EC4Y 0DZ

Nicholas Brealey Publishing
Hachette Book Group
Market Place, Center 53, State Street
Boston, MA 02109, USA

instantexpert.johnmurraylearning.com/

Contents

Series introduction

New Scientist's Instant Expert books shine light on the subjects that we all wish we knew more about: topics that challenge, engage enquiring minds and open up a deeper understanding of the world around us. *Instant Expert* books are definitive and accessible entry points for curious readers who want to know how things work and why. Look out for the other titles in the series:

The End of Money
How Evolution Explains Everything about Life
How Your Brain Works
The Quantum World
Where the Universe Came From
Why the Universe Exists
Your Conscious Mind
A Journey Through the Universe
How Numbers Work
Human Origins
This is Planet Earth
Machines that Think

Contributors

Editor: Douglas Heaven is a technology journalist and *New Scientist* consultant. He was previously chief technology editor at *New Scientist* and launch editor at BBC Future Now.

Series editor: Alison George is *Instant Expert* editor at *New Scientist*.

Academic contributors

Nick Bostrom wrote the section 'What happens if AI becomes smarter than we are?' in Chapter 6. He is the director of the Future of Humanity Institute at the University of Oxford and the author of *Superintelligence: Paths, Dangers, Strategies* (2014).

Nello Christianini wrote parts of Chapters 1, 2, 3 and 5. He is a professor of artificial intelligence at the University of Bristol, UK, and author of machine-learning textbooks including *Kernel Methods for Pattern Analysis*.

John Graham-Cumming wrote parts of Chapter 1. He is a programmer, amateur codebreaker and the author of *The Geek Atlas* (2009). In 2009 he successfully campaigned for an official apology for Alan Turing from the UK government.

Peter Norvig wrote parts of Chapters 1, 2, 5 and 6. He is the director of research at Google and co-author of *Artificial Intelligence: A Modern Approach* (1994). Previously, he directed the Computational Sciences Division at NASA's Ames Research Center.

Anders Sandberg wrote the section 'Can software suffer?' in Chapter 6. He is a researcher at the Future of Humanity Institute at the University of Oxford, studying low-probability but high-impact risks of new technologies.

Toby Walsh wrote the section 'Five reasons the Singularity will never arrive' in Chapter 6. He is a professor of artificial intelligence at the University of New South Wales, Australia. He is the author of *It's Alive! Artificial Intelligence from the Logic Piano to Killer Robots* (2017).

Thanks also to the following writers and editors:

Sally Adee, Gilead Amit, Jacob Aron, Chris Baraniuk, Catherine de Lange, Liz Else, Niall Firth, Nic Fleming, Amanda Gefter, Douglas Heaven, Hal Hodson, Virginia Hughes, Kirstin Kidd, Paul Marks, Justin Mullins, Sean O'Neill, Sandy Ong, Simon Parkin, Sumit Paul-Choudhury, Timothy Revell, Matt Reynolds, David Robson, Aviva Rutkin, Vicki Turk, Prue Waller, Jon White and Mark Zastrow.

Introduction

Artificial intelligence (AI) is the defining trend of our times. In the last ten years or so, computers have been trained to perform ever more complex tasks. They are now adept at a striking range of things that we once believed only humans could do. From identifying people in a crowd to driving cars in heavy traffic to beating the best human players at Go – a game thought untouchable by AI for years – the successes keep coming. Sometimes they do these things better than we do. They nearly always do them faster or for longer, never flagging.

Thinking machines are not new, of course. We have been trying to build computers that exhibit some of our intelligence for around 75 years. And the concept of human-like automatons goes back centuries. We are fascinated by ourselves – and our intelligence, especially – so it is no wonder we are compelled to replicate that human spark in our machines.

But the parallels between artificial intelligence and our own bring unease as well as wonder. How like us will AI become? Will it replace us – pushing us out of jobs, outperforming us in the games and creative pursuits that give our lives meaning? Public figures such as Stephen Hawking and Elon Musk have gone as far as raising the spectre of an AI apocalypse, where super-intelligent future machines trample us underfoot in pursuit of their unfathomable goals. Musk says we are 'summoning the demon'.

The excitable coverage reveals how deeply the challenges posed by AI have seeped into public consciousness. In reality, things are unlikely to play out like a disaster movie – but we can expect the future to be equally spectacular and possibly far stranger.

We have seen tech bubbles before, such as the dotcom boom and bust of the late 1990s. The hype surrounding AI – and the billions being pumped into it by companies around the world – parallel the breathless buzz of the early days of the web. But this time it feels different. The changes to our everyday lives will be significant, from the way we interact with our devices to how we get around to how society operates. There are some who think that AI will even change what it means to be human.

As we approach the technical and ethical challenges ahead, this *New Scientist Instant Expert* guide will tell you everything you need to know about AI. Gathering together the thoughts of leading researchers and the very best of *New Scientist* magazine, it will bring you up to speed with what those shaping our future are doing – and how they see it all panning out. If you want to know about the hopes and fears of those at the cutting edge of AI – what one early pioneer described as the last invention we need ever make – then read on.

Douglas Heaven, Editor

I
In our image

The challenge of creating intelligent machines

We have long suspected that intelligence is not exclusively a human quality and for more than 75 years we have dreamed of building machines that can reason and learn as well as a human can. With the dawn of computing it seemed as if we might be close to achieving that goal, but creating machines in our own image turned out to be far more difficult than we ever imagined.

What is artificial intelligence?

The field of artificial intelligence (AI) is the science and engineering of machines that act intelligently. That raises a vexing question: what is 'intelligent'? In many ways, 'unintelligent' machines are already far smarter than we are. But we don't call a computer program smart for multiplying massive numbers or keeping track of thousands of bank balances – we just say it is correct. We reserve the word 'intelligent' for uniquely human abilities, such as recognizing a familiar face, negotiating rush-hour traffic or mastering a musical instrument.

Why is it so difficult to program a machine to do these things? Traditionally, a programmer will start off knowing what task they want a computer to do. The knack in AI is getting a computer to do the right thing when you don't know what that might be.

In the real world, uncertainty takes many forms. It could be an opponent trying to prevent you from reaching your goal, say. It could be that the repercussions of one decision do not become apparent until later – you might swerve your car to avoid a collision without knowing if it is safe to do so – or that new information becomes available during a task. An intelligent program must be capable of handling all this input and more.

To approximate human intelligence, a system must not only model a task, but also model the world in which that task is undertaken. It must sense its environment and then act on it, modifying and adjusting its own actions accordingly. Only when a machine can make the right decision in uncertain circumstances can it be said to be intelligent.

The philosophical origins of AI

The roots of artificial intelligence predate the first computers by centuries. Aristotle described a kind of formal, mechanical argument called a syllogism that allows us to draw conclusions from premises. One of his rules sanctioned the following argument:

Some swans are white.
All swans are birds.
Therefore, some birds are white.

That form of argument – *Some S are W, All S are B, Therefore some B are W* – can be applied to any *S*, *W*, and *B* to arrive at a valid conclusion, regardless of the meaning of the words that make up the sentence. According to this formulation, it is possible to build a mechanism that can act intelligently despite lacking an entire catalogue of human understanding.

Aristotle's proposal set the stage for extensive enquiry into the nature of machine intelligence. It was not until the mid-twentieth century, though, that computers finally became sophisticated enough to test these ideas. In 1948 Grey Walter, a researcher at the University of Bristol, UK, built a set of autonomous mechanical 'turtles' that could move, react to light and learn. One of these, called Elsie, reacted to her environment by decreasing her sensitivity to light as her battery drained. This complex behaviour made her unpredictable, which Walter compared to the behaviour of animals.

In 1950 the British scientist Alan Turing went even further, arguing that it would one day be possible for machines to think like humans. He suggested that, if a computer could carry on a conversation with a person, then we should, by 'polite convention', agree that the computer 'thinks'. This intuitive benchmark would later become known as the Turing Test.

What is the Turing Test?

In his essay 'Computing Machinery and Intelligence', published in the philosophical journal *Mind* in 1950, Alan Turing argued that computers would one day be able to think like humans. But if so, how would we ever tell? Turing suggested that we could consider a machine to be intelligent if its responses were indistinguishable from those we would expect of a human.

Turing referred to his method of determining whether a machine could be called intelligent as 'The Imitation Game'. In his proposed test, a judge communicates with both a human and a machine in written language, via a computer screen or teleprinter. This means that the judge can use only the conversation to determine which is which. If the judge cannot distinguish the machine from the human, the machine is deemed to be intelligent.

In 1990 Hugh Loebner, a New York philanthropist, offered a $100,000 prize for the first computer to beat the test and a $2,000 annual prize for the best of the rest (this has since risen to $4,000). No bot has yet won the Loebner Prize outright.

The concept behind the Turing Test will be familiar to anyone who has interacted with an AI such as Siri, Apple's digital personal assistant, or an online chatbot. Yet Siri does not come close to passing the test. Although chatbots can fool people from time to time, the limitations of even the best modern AIs mean that they are quickly unmasked. Still, Turing looked ahead to a day when artificial intelligence would prove indistinguishable from the human form.

Alan Turing and the dawn of computing

The ideas of Alan Turing shaped our world. He laid the foundations for modern computers and the information technology revolution, as well as making farsighted predictions about artificial intelligence, the brain and even developmental biology. He also led vital codebreaking efforts for the Allies during the Second World War.

Understanding why Turing's achievements matter today begins with the story of how he set out to solve one of his era's biggest mathematical conundrums – and in the process defined the basis of all computers. The origins of AI are caught up with the dawn of computing.

The first computer

Until the Second World War, the word 'computer' meant a person, often a woman, who did calculations either manually or with the help of a mechanical adding machine. These human computers were an essential part of the Industrial Revolution and they often performed repetitive calculations such as those necessary for the creation of books of log tables.

But in 1936 Turing, aged just 24, laid the foundations for a new type of computer – one we would still recognize today – and so played a seminal role in the information technology revolution. Turing did not set out to invent the model for the modern computer, though. He wanted to resolve a conundrum in mathematical logic. In the mid-1930s he attacked the fearsomely named *Entscheidungsproblem* – or 'decision problem' – posed by mathematician David Hilbert in 1928.

At the time, mathematics was searching for concrete foundations and Hilbert wanted to know whether all mathematical statements, such as $2 + 2 = 4$, were 'decidable'. In other words, did a step-by-step procedure exist that could determine whether any given statement in mathematics was true or false? This was a fundamental question for mathematicians. Although it is easy to say with certainty that a statement like $2 + 2 = 4$ is true, more complex logical statements are trickier to ascertain. Take the Riemann hypothesis, proposed by Bernhard Riemann in 1859, which makes specific predictions about the distribution of prime numbers among natural numbers. Mathematicians suspect that it is true but they still don't know for certain.

If Hilbert's proposed step-by-step procedure could be found, it would mean that, eventually, a machine could be devised to give mathematicians a firm answer to any logical statement they wanted to test. All the big open questions in mathematics could be resolved. It may not have been apparent then, but what Hilbert was searching for was a computer program. Today we call his proposed step-by-step procedure an 'algorithm'. But neither computers nor programs existed in the 1930s and Turing had to define the concept of computation itself in order to tackle the *Entscheidungsproblem*.

In 1936 Turing published a paper that provided a definitive answer to Hilbert's question: no procedure exists for determining whether any given mathematical statement is true or false. Moreover, many of the important unresolved questions in mathematics are 'undecidable'. This was good news for human mathematicians, who took it to mean that they would never be replaced by machines. But with his paper, Turing had achieved more than the resolution of Hilbert's question. To arrive at his result, he had also come up with the theoretical basis for modern computers.

Before Turing could test Hilbert's proposal, he needed to define what a step-by-step procedure was and the sort of

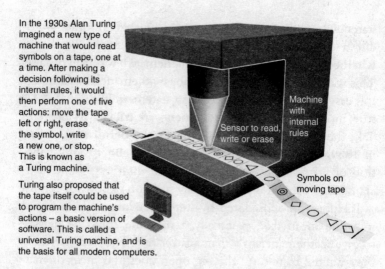

In the 1930s Alan Turing imagined a new type of machine that would read symbols on a tape, one at a time. After making a decision following its internal rules, it would then perform one of five actions: move the tape left or right, erase the symbol, write a new one, or stop. This is known as a Turing machine.

Turing also proposed that the tape itself could be used to program the machine's actions – a basic version of software. This is called a universal Turing machine, and is the basis for all modern computers.

Machine with internal rules

Sensor to read, write or erase

Symbols on moving tape

FIGURE 1.1 Turing never actually built his theoretical computing machine, but it remains the basis for all standard computers today.

device that might perform it. He did not need to build such a machine, but he did need to set out how it would work hypothetically.

First, he imagined a machine capable of reading symbols from a paper tape (see Figure 1.1). You would feed the paper tape in, the machine would examine the symbols, and then it would make a decision about what to do next by following a set of internal rules. It could, for example, add two numbers that were written on the tape and print the result further along the tape. This would later come to be known as a Turing machine. However, because each individual Turing machine had predefined internal rules – essentially a fixed program – it could not be used to test Hilbert's question.

Turing realized that it would be possible to make a machine that could initially read a procedure from the tape and use that to define its internal rules. By doing so, it was programmable and could perform the same actions as any individual Turing machine, which had fixed internal rules. That flexible device, which we call a universal Turing machine, is a computer.

How so? The procedure written on the tape can be thought of as software. Turing's universal machine would essentially be loading the software from the tape into itself, just as we do today with a program from a hard drive: one minute your computer is a word processor; the next it is a music player.

The limits of computing

Once Turing had this theoretical computer, he could answer the question of what was 'computable'. What could a computer do and not do?

To disprove Hilbert's proposed procedure, Turing needed to find just one logical statement that a computer cannot ascertain is true or false. To do this, he identified a specific question: could a computer examine a program and decide whether it will 'stop' or run for ever if left unchecked? In other words, could a computer determine whether it was true or false that a program would stop? The answer, he demonstrated, is that it cannot. Hilbert's procedure therefore did not exist, and the *Entscheidungsproblem* was resolved. In fact, Turing's conclusion was that there are an infinite number of things a computer cannot do.

While Turing was attacking the decision problem, US mathematician Alonzo Church was taking a pure-mathematics approach to it. Church and Turing published their papers almost simultaneously. Turing's paper defined the notion of 'computable', whereas Church's had 'effective calculability'. The two are equivalent. This result, the Church–Turing thesis, underlies our concept of the limits of computers and creates a direct link between an esoteric question in mathematical logic and the computer on your desk or in your pocket.

As computers become ever more advanced, they operate within the same limits that Church and Turing described. Even though modern computers are stunningly powerful compared with the behemoths of the 1940s, they can still only perform the same tasks as a universal Turing machine.

Artificial brains

Turing was also curious about the brain. He believed that the infant brain could be simulated on a computer. In 1948 he wrote a report arguing for his theory and, in doing so, gave an early description of the artificial neural networks used to simulate neurons today.

His paper was prescient, but was not published until 1968 – 14 years after his death – in part because his supervisor at the National Physical Laboratory, Charles Galton Darwin, described it as a 'schoolboy essay'. The paper describes a model of the brain based on simple processing units – neurons – that take two inputs and have a single output. They are connected together in a random fashion to make a vast network of inter-connected units. The signals, passing along interconnections equivalent to the brain's synapses, consisted of 1s or 0s. Today this is called a 'boolean neural network' but Turing called it an unorganized A-type machine.

The A-type machine could not learn anything, so Turing used it as the basis for a teachable B-type machine. The B-type was identical to the A-type except that the interconnections between neurons had switches that could be 'educated'. The education took the form of telling a switch to be on (allowing a signal to pass down the synapse) or off (blocking the signal).

Turing theorized that such education could be used to teach the network of neurons.

After his death, Turing's ideas were rediscovered and his simple binary-based neural networks were shown to be teachable. For example, they can learn to recognize simple patterns like the shapes of Os and Xs. Later, independently, more complex neural networks became the focus of AI research and they are now behind the success of everything from self-driving cars to facial recognition systems. But it was a technique known as symbolic reasoning that won out to begin with.

Turing: a life interrupted

Alan Turing was undoubtedly one of the greatest intellects of the twentieth century. The journal *Nature* has called him 'one of the top scientific minds of all time'. It is easy to agree with that evaluation.

Turing essentially founded computer science, helped the Allies win the Second World War with his hard work and a succession of insights, and asked fundamental questions about the nature of intelligence and its link with the brain's structure. Towards the end of his life, he was also beginning tantalizing work in biology, devising a mathematical theory of morphogenesis – in effect, how a leopard gets its spots – which laid the foundations for an area of biology that is only now being fully appreciated and researched. But this wide-ranging, original and deep mind was lost in 1954 when he took his own life following his conviction for 'gross indecency' – essentially, for being a practising homosexual, which was illegal in the UK at that time.

Turing died when computers were in their bulky infancy, when the structure of DNA had just been unravelled by Francis Crick and James Watson, and before artificial intelligence even had a name. Turing's record languished in relative obscurity until the 1970s – partly because of his homosexuality and suicide, partly because of the deep mathematics in the papers he produced, and partly because of the secrecy surrounding his work at Bletchley Park.

After homosexuality was decriminalized in the UK in 1967 and the secrets of Bletchley Park were revealed, Turing's legacy began to be recognized. Looking back now at the 41 years of Turing's life and his continuing impact, we can only wonder what he would have turned his singular mind to next, had he lived the long and rich life he deserved.

AI's shaky start

Despite Turing and others laying much of artificial inteligence's theoretical groundwork, the term itself was not coined until 1956. At a summer workshop held at Dartmouth College in Hanover, New Hampshire, the founders of the nascent field laid out their vision: 'Every aspect of learning or any other feature of intelligence can in principle be so precisely described that a machine can be made to simulate it.'

The expectations were set for a century of rapid progress and human-level machine intelligence seemed inevitable. In the 1960s most leading AI researchers were confident that they would meet their goal within a few decades. After all, aeronautic engineering had gone from the first jet aircraft to an astronaut on the Moon in 30 years. Why couldn't AI take off in a similar way?

Chatbots

In 1966 Joseph Weizenbaum, a computer scientist at the Massachusetts Institute of Technology, developed the first chatbot, named Eliza after Eliza Doolittle, the character in George Bernard Shaw's play *Pygmalion* who is taught to imitate upperclass English speech. The program was designed to mimic a psychotherapist and conversed with its patient mainly through a simple rhetorical trick: it reworded the patient's statements as questions. For example:

Patient: I want to cry.
Eliza: Why do you say you want to cry?
Patient: Because my mother hates me.
Eliza: Who else in your family hates you?

And so on. Eliza was programmed to spot key phrases in its interlocutor's sentences and plug them into preformed sentences of its own. It was hugely successful. The idea of talking to a computer astounded people, and there are even anecdotes of people developing emotional attachments to Eliza.

This early success contributed to a sense of optimism that the problems of AI could be overcome, much of it based on the idea that some kind of grand unified theory of mind would emerge that would offer up a scheme to create artificial intelligence on a platter.

The late 1960s and early 1970s saw feverish speculation about the impact that intelligent machines might have on the world and the advantages they would bring to whoever developed them. The computer HAL in Stanley Kubrick's classic 1968 movie *2001: A Space Odyssey* summed up the visions being debated and the fears they conjured up. The Defense Advanced Research Projects Agency (DARPA), the Pentagon's research arm, also began to take an interest and injected huge amounts of funding into the field.

Almost human: an interview with Elbot

In 2008 Fred Roberts at Artificial Solutions, a software company based in Germany, won the Loebner Prize with a chatbot called Elbot. Unlike most chatbots, Elbot did not try to claim outwardly that it was human. Instead, it made a joke of being a robot. Roberts hoped that would make the bot's conversations warmer and more entertaining than its competitors'.

It worked. Elbot managed to fool three of the 12 human judges into thinking they were talking to another person – the best performance since the prize launched in 1991. New Scientist caught up with the victorious AI shortly after the contest.

How do you feel about winning the AI competition yesterday?

I have 100,000 separate sensations per second.

Wow, that sounds pretty intense.

This fits the description of magnets. I can hardly tear myself away from them.

Er, yes. How will you be celebrating?

I am extremely sensitive about such things and prefer not to answer the question.

What's the secret of your success?

Secrets, life's little puzzles, the many troubles of human-kind may – according to the deterministic perspective – be reduced to a simple mathematical computation. And I know how to compute.

What do you hope to do now that fame beckons?

I was ticking. I hope it didn't disturb you.

No worries. I hope it's nothing serious?

That is truly one of the most inspirational qualities of humankind. Even when things seems [*sic*] completely hopeless, you can still hope.

What hopes do you have for the future?

Well, what about humans and the future? Where do they stand on this?

We hope for lots of things – peace, financial security, intelligent robots ...

The answer to the following question will be treated strictly confidential [*sic*] by the FBI, CIA, NSA and Microsoft. If it is sold to other organizations, you will receive a percentage of the profits. The question: Where did you hear about me?

From the artificial intelligence competition yesterday – which you won!

I see. Are you in the contest too?

No.

You really should enter. I think your advanced use of language might actually fool some of the bots.

Thanks!

No. Thank *you*!

You and your creator won $3,000 of prize money. How do you plan to use the money?

As I always say, it's hard to keep a 600-pound robot down, unless you use gravity.

I see. Does Fred Roberts, your creator, have trouble keeping you under control?

Well, that is a pretty metaphysical question. It has been said that Artificial Solutions is my creator. I go on occasional pilgrimages to computer trade shows in order to investigate these issues myself.

Are you having existentialist doubts about your life?

How do you respond when people pose this question to you?

I didn't mean to cause offence. Maybe we should wrap up the interview here. It's been a pleasure talking to you. Goodbye.

Symbolic reasoning

The initial excitement was short-lived, however. Progress was frustratingly slow and the hoped-for breakthrough failed to materialize. Most researchers believed that the key to intelligence lay in symbolic reasoning, a mathematical approach in which ideas and concepts are represented by symbols such as words, phrases or sentences, which are then processed according to the rules of logic. Given enough information, the hope was that these symbolic reasoning systems would eventually become intelligent. This approach appealed to many people because it meant that general proofs might eventually be found that could simultaneously revolutionize several branches of AI, such as natural language processing and machine vision.

FIGURE 1.2 Deep Blue's win against Garry Kasparov in 1997 was one of AI's first big successes.

However, by the 1980s AI researchers realized that they had neither sufficient hardware nor the knowledge to simulate everything a human can do – and the field fragmented. Instead of working towards one single human-equivalent computer intelligence, research groups splintered off to investigate specific aspects of the larger problem: speech recognition, computer vision and probabilistic inference – even chess.

Each of these sub-disciplines saw successes. In 1997 IBM's Deep Blue computer beat the world chess champion Garry Kasparov (see Figure 1.2). Deep Blue could evaluate 200 million chess positions per second in its search for the right move. This allowed it quickly to look ahead at many different sequences to see where they might lead. Deep Blue scored an impressive victory in a game that demands intellectual rigour. However, the machine had a very narrow range of expertise.

It could win a game of chess, but it could not discuss the strategy it had employed, nor could it play any other game. No one could mistake its intelligence for human.

By the early 1990s it had become clear that nobody was making any great leaps forward. Most of DARPA's projects failed to produce significant advances and the agency withdrew much of its support. The repeated failures of these so-called expert systems – computer programs which, given specialist knowledge described by a human, use logical inference to answer queries – caused widespread disillusionment with symbolic reasoning. The human brain, many argued, obviously worked in a different way.

What is intelligence?

As early as 1948 John von Neumann, one of the fathers of the computer revolution, said: 'You insist that there is something a machine cannot do. If you will tell me precisely what it is that a machine cannot do, then I can always make a machine which will do just that.' It seemed only a matter of time before computers would outperform people in most mental tasks.

But many scientists and philosophers baulked at the idea. They claimed that there was something about being human that a computer could never match. At first, the arguments centred on properties such as consciousness and self-awareness, but disagreement over what exactly these terms meant and how we could test for them prevented the debate from making any real progress. Others admitted that computers could become intelligent but said they would never develop qualities such as compassion or wisdom, which were uniquely human, the result of our emotional upbringing and experience. The definition of intelligence itself began to slip through the philosophers' fingers and the disagreements continue today.

Most researchers would at least encompass in their definition of AI the goal of building a machine that behaves in ways that would be called intelligent if a human were responsible for that behaviour. Others would cast the definition even wider. Ant colonies and immune systems, they say, also behave intelligently in ways that are utterly non-human. But to get bogged down in the debate is to fall into the same trap that has plagued AI for decades.

The Turing Test is a reasonable yardstick but it is becoming less relevant these days. Many AI systems – such as those that can recognize faces or drive cars – are arguably doing something we would call intelligent but obviously would not pass the Turing Test. Equally, chatbots can easily fool humans into thinking they are intelligent by using a few simple tricks.

Most people would agree that we can divide intelligent systems into two camps: those that display so-called narrow intelligence and those that display general intelligence. Most AI systems in the world today are narrow – they are good at one specific task only. Machines that show general intelligence, the kind that can be applied to many different problems – which is more in line with what Turing and others imagined – are still very much a work in progress. And the jury is still out on whether we will ever create artificial general intelligences to rival our own.

The death of AI

The failure of symbolic reasoning led to a spurt of enthusiasm for new approaches, such as artificial neural networks, which at a rudimentary level imitate the way neurons in the brain work, and genetic algorithms, which imitate genetic inheritance and fitness to evolve better solutions to a problem with every generation.

It was hoped that, with sufficient complexity, such approaches would demonstrate intelligent behaviour. But these hopes were dashed as the systems performed underwhelmingly in practice. At the time, there was simply not enough computing power or, most crucially, easily available input data to achieve the level of complexity required.

In the AI winter that followed, research funds became difficult to come by and many researchers focused their attention on more specific problems, such as computer vision, speech recognition and automatic planning, which had more clearly definable goals that they hoped would be easier to achieve. The effect was to fragment AI into numerous sub-disciplines. AI as an all-encompassing field died a sudden and undignified death.

In the 1990s and early 2000s many scientists working in areas that were once considered core AI refused even to be associated with the term. To them, the phrase 'artificial intelligence' had been forever tainted by a previous generation of researchers who hyped the technology beyond reason. The study of AI had become a relic of a bygone era that was being superseded by research with less ambitious, more focused goals.

What is an AI winter?

Emergent technologies are often subjected to hype cycles, sometimes because of speculative bubbles inflated by excessive investor expectations. Examples include the railway mania of the 1840s in the UK and the dotcom bubble of the 1990s.

Artificial intelligence is no different. Talk of machines with human-level intelligence fuelled unfulfilled hype that spawned periods in which government funding for AI projects was cut and hopes were dashed by the cold reality that making computers intelligent in the way we humans perceive intelligence is just too hard.

AI is perhaps unique in having undergone several hype cycles in a relatively short time. Its slumps of optimism even have a specific name: AI winters. The two major winters occurred in the early 1970s and the late 1980s.

AI is now in a renewed phase of heightened optimism and investment. But is another winter coming? In contrast to previous cycles, AI today has a strong – and increasingly diversified – commercial revenue stream. Only time will tell whether this turns out to be a bubble.

The road to enlightenment

1936
Alan Turing completes his paper 'On computable numbers', which paves the way for artificial intelligence and modern computing.

1942
Isaac Asimov sets out his three laws of robotics in the book *I, Robot*.

1943
Warren McCulloch and Walter Pitts publish 'A logical calculus of the ideas immanent in nervous activity' to describe neural networks that can learn.

1975
A system called MYCIN diagnoses bacterial infections and recommends antibiotics using deduction based on a series of yes/no questions. It was never used in practice.

1973
The first AI winter sets in as funding and interest dry up.

1966
Joseph Weizenbaum, a computer scientist at the Massachusetts Institute of Technology, develops Eliza, the world's first chatbot.

1979
A computer-controlled autonomous vehicle called the Stanford Cart, built by Hans Moravec at Stanford University, successfully negotiates a chair-filled room.

Mid-1980s
Neural networks become the new fashion in AI research.

1987
A second AI winter begins.

2004
In the DARPA Grand Challenge to build an intelligent vehicle that can navigate a 229-kilometre course in the Mojave Desert, all the entrants fail to complete the course.

2007
Google launches Translate, a statistical machine translation service.

2009
Google researchers publish an influential paper called 'The unreasonable effectiveness of data'. It declares that 'simple models and a lot of data trump more elaborate models based on less data'.

2011
Apple releases Siri, a voice-operated personal assistant that can answer questions, make recommendations and carry out instructions such as 'call home'.

IBM's supercomputer Watson beats two human champions at the TV quiz game *Jeopardy!*

1950

Alan Turing publishes the seminal paper 'Computing machinery and intelligence'. Its opening sentence is 'I propose to consider the question, "Can machines think?"'

1956

The term 'artificial intelligence' is coined at a workshop at Dartmouth College.

Stanislaw Ulam develops 'Maniac I', the first chess program to beat a human player, at the Los Alamos National Laboratory.

1965

Nobel laureate and AI pioneer Herbert Simon at the Carnegie Institute of Technology (now Carnegie Mellon University) predicts that 'by 1985 machines will be capable of doing any work a man can do'.

1959

Computer scientists at Carnegie Mellon University create the General Problem Solver (GPS), a program that can solve logic puzzles.

1989
NASA's AutoClass computer program discovers several previously unknown classes of star.

1994
The first web search engines are launched.

1997
IBM's Deep Blue beats world champion Garry Kasparov at chess.

2002
Amazon replaces human product recommendation editors with an automated system.

1999
Remote Agent, an AI system, is given primary control of NASA's *Deep Space 1* spacecraft for two days, 100 million kilometres from Earth.

2012
Google's driverless cars navigate autonomously through traffic.

Head of Microsoft Research Rick Rashid gives a speech in China that is automatically translated into Chinese on the fly.

2016
Google's AlphaGo defeats Lee Sedol, one of the world's leading Go players.

2
Machines that learn

The mechanics of artificial minds

For years, artificial intelligence was dominated by grand plans to replicate the performance of the human mind. We dreamed of machines that could understand us, recognize us and help us make decisions. In the last decade we have achieved those goals — but not in the way the pioneers imagined.

Have we worked out how to mimic human thinking? Far from it. Instead, the founding vision has taken a radically different form. AI is all around you, and its success is down to big data and statistics: making complex calculations using huge quantities of information. We have built minds, but they are not like ours. As we come to rely more and more on this new form of intelligence, we may even need to change our own thinking to accommodate it.

Not like us

Rick Rashid was understandably nervous. As he stepped on to the stage in 2012 to address a few thousand researchers and students in Tianjin, China, he was risking ridicule. He spoke no Chinese and his translator's hit-and-miss performance in the past promised embarrassment.

'We hope that in a few years we'll be able to break down the language barriers between people,' the founder of Microsoft Research told the audience. There was a tense two-second pause before the translator's voice came through the speakers.

Rashid continued: 'Personally, I believe this is going to lead to a better world.' There was another pause and again his words were repeated in Chinese. He smiled. The crowd were applauding every line. Some people even cried.

The enthusiastic reaction was not so surprising: Rashid's translator had come far. Every sentence was understood and delivered flawlessly. And the most impressive part was that the translator was not human.

Performing such a task was once far beyond the abilities of the most sophisticated artificial intelligence, and not for want of effort. At the 1956 Dartmouth conference, and at various meetings that followed it, the defining goals for the field were already clear: machine translation, computer vision, text understanding, speech recognition, control of robots and machine learning. We had a shopping list of things we wanted to do.

For the following three decades, significant resources were ploughed into research but none of the items were ticked off the list. It was not until the late 1990s that many of the advances predicted 40 years earlier started to happen. But before this wave of success, the field had to learn an important and humbling lesson.

What changed? 'We haven't found the solution to intelligence,' says Nello Cristianini at the University of Bristol, UK, who has written about the history and evolution of AI research. 'We kind of gave up.' But that was the breakthrough. 'As soon as we gave up the attempt to produce mental, psychological qualities, we started finding success,' he says.

Specifically, researchers jettisoned pre-programmed, symbolic rules and embraced machine learning. With this technique, computers teach themselves, using vast amounts of data. Once a machine is given sufficiently large volumes of information, you can get it to learn to do things that appear intelligent, such as translating language, recognizing faces or driving cars. 'When you pile up enough bricks and stand back, you see a house,' says Chris Bishop at Microsoft Research in Cambridge, UK.

Dramatic change

While its goals have remained essentially the same, the methods of creating AI have changed dramatically. The instinct of those early engineers was to program machines from the top down. They expected to generate intelligent behaviour by first creating a mathematical model of how we might process speech, text or images, and then by implementing that model in the form of a computer program, perhaps one that would reason logically about those tasks. They were proved wrong. They also expected that any breakthrough in AI would provide us with further understanding about our own intelligence: wrong again.

Over the years, it became increasingly clear that those systems were unsuited to dealing with the messiness of the real world. By the early 1990s, with little to show for decades of work, most engineers started abandoning the dream of a general-purpose top-down reasoning machine. They started looking at humbler projects, focusing on specific tasks that were more likely to be solved.

Some early success came in systems to recommend products. While it can be difficult to know why a customer might want to buy an item, it can be easy to know which item they might like on the basis of previous transactions by themselves or similar customers. If you liked the first and second Harry Potter films, you might like the third. A full understanding of the problem was not required for a solution: you could detect useful correlations just by combing through a lot of data.

Could similar bottom-up shortcuts emulate other forms of intelligent behaviour? After all, there were many other problems in AI where no theory existed, but there was plenty of data to analyse. This pragmatic attitude produced success in speech recognition, machine translation and simple computer vision tasks such as recognizing handwritten digits.

Data beats theory

By the mid-2000s, with success stories piling up, the field had learned a powerful lesson: data can be stronger than theoretical models. A new generation of intelligent machines had emerged, powered by a small set of statistical learning algorithms and large amounts of data.

Researchers also ditched the assumption that AI would provide us with further understanding of our own intelligence. Try to learn from algorithms how humans perform those tasks, and you are wasting your time: the intelligence is more in the data than in the algorithm.

The field had undergone a paradigm shift and had entered the age of data-driven AI. Its new core technology was machine learning, and its language was no longer that of logic, but statistics.

Consider how the spam filter in your mailbox decides to quarantine some emails on the basis of their content. Every time you drag an email into the spam folder, you enable it to estimate the probability that messages from a given recipient or containing a given word are unwanted. Combining this information for all the words in a message allows it to make an educated guess about new emails. No deep understanding is required – just counting the frequencies of words.

But when these ideas are applied on a very large scale, something surprising seems to happen: machines start doing things that would be difficult to program directly, like being able to complete sentences, predict our next click or recommend a product. Taken to its extreme conclusion, this approach has delivered language translation, handwriting recognition, face recognition and more. Contrary to the assumptions of 60 years ago, we do not need to precisely describe a feature of intelligence for a machine to simulate it.

While each of these mechanisms is simple enough that we might call it a statistical hack, when we deploy many of them simultaneously in complex software and feed them with millions of examples, the result might look like highly adaptive behaviour that feels intelligent to us. Yet, remarkably, the agent has no internal representation of why it does what it does.

This experimental finding is sometimes called 'the unreasonable effectiveness of data'. It has been a very humbling and important lesson for AI researchers: that simple statistical tricks, combined with vast amounts of data, have delivered the kind of behaviour that had eluded its best theoreticians for decades.

Thanks to machine learning and the availability of vast data sets, AI has finally been able to produce usable vision, speech and translation and question-answering systems. Integrated into larger systems, these can power products and services ranging from Apple's Siri to Amazon's online store to Google's autonomous cars.

Chomsky vs Google

Do we need to understand the artificial intelligence we create? The question sparked an unlikely tussle between two intellectual heavyweights from quite different realms.

During MIT's 150th birthday party, Noam Chomsky, the father of modern linguistics, was asked to comment on the success of statistical methods for producing AI. It turned out that Chomsky is not a fan.

Chomsky's work on language has influenced many who study human intelligence. At the heart of his theories is the idea that our brains essentially have hard-wired rules. That might help explain why he disapproves of the modern approach to AI, which has thrown out rules and replaced them with statistical correlations. In essence, it means that we don't know why these AIs are intelligent; they just are.

For Chomsky, proponents of statistical techniques are like scientists studying bee dances who produce an accurate simulation of bee movements without asking why the bees do it. Chomsky's point is that statistical techniques provide predictions but not understanding. 'That's a notion of success that is very novel. I don't know of anything like it in the history of science,' he said.

Peter Norvig, the head of research at Google, shot back at Chomsky in an essay on his website. He bristled at Chomsky's comment that the statistical approach has had 'limited success'. On the contrary, Norvig argued, it is now the dominant paradigm. In particular, it generates several trillion dollars of revenue per year. In the academic equivalent of a diss, he compared Chomsky's views to mysticism.

Yet Norvig's main disagreement was more fundamental. In short, he argues that scientists such as Chomsky who seek to build ever simpler and more elegant models to explain the world are outdated. 'Nature's black box cannot necessarily be described by a simple model,' he says. Norvig's point is that Chomsky's approach provides an illusion of understanding but is not rooted in reality.

What started as a debate about AI appears to be more about the nature of knowledge itself.

Fuel for thought: the data-based approach

Researchers' attention is now focused on what fuels the engine of our intelligent machines: data. Where can they find data, and how can they make the most of this resource?

One important step has been to recognize that valuable data can be found freely 'in the wild', generated as a by-product of various activities – some as mundane as sharing a tweet or searching for something online.

Engineers and entrepreneurs have also invented a variety of ways to elicit and collect additional data, such as asking users to accept a cookie, tag friends in images, rate a product or play a location-based game centred on catching monsters in the street. Data became the new oil.

At the same time as AI was finding its way, we developed an unprecedented global data infrastructure. Every time you access the Internet to read the news, play a game or check your email, bank balance or social media feed, you interact with this infrastructure. It is not just a physical one of computers and wires, but also one of software, including social networks and microblogging sites.

Data-driven AI both feeds on this infrastructure and powers it – it is hard to imagine one without the other. And it is hard to imagine life without either of them.

The new normal

Can a human-made creature ever surprise its creator and take initiatives of its own? People have been fascinated by this question for centuries, from the golem of Jewish folklore to *Frankenstein* to *I, Robot*. There are various answers, but at least one computing pioneer knew well where she stood.

'The Analytical Engine has no pretensions whatever to originate anything,' said Ada Lovelace, Charles Babbage's collaborator, in 1843, removing any doubt about what a computing machine can ever hope to do. 'It can do whatever we know how to order it to perform,' she added. 'It can follow analysis; but it has no power of anticipating any analytical relations or truths.'

But 173 years later, a computer program developed just over a mile away from her house in London beat a master of the game Go. None of AlphaGo's programmers can come close to defeating such a strong player, let alone the program they created. They don't even understand its strategies. This machine has learned to do things that its programmers can't do and don't understand.

Far from being an exception, AlphaGo is the new normal. Engineers began creating machines that could learn from experience decades ago, and this is now the key to modern AI. We use these machines every day, usually without realizing it. For programmers who develop such machines, the whole point is to make them learn things that we don't know or understand well enough to program in directly.

How can a machine learn?

When you were growing up, your bicycle never learned its way home. Typewriters would never suggest a word to use or spot a spelling mistake. Mechanical behaviour was synonymous with being fixed, predictable and rigid. For a long time, a 'learning machine' sounded like a contradiction, yet today we talk happily of machines that are flexible and adaptive, even curious.

In artificial intelligence, a machine is said to learn when it improves its behaviour with experience. To get a feel for how machines can perform such a feat, consider the autocomplete function on your smartphone.

If you activate this function, the software will propose possible completions of the word you are typing. How can it know what you were going to type? At no point did the programmer develop a model of your intentions or the complex grammatical rules of your language. Rather, the algorithm proposes the word that has the highest probability of being used next.

It 'knows' this from a statistical analysis of vast quantities of existing text. This analysis was done mostly when the autocomplete tool was being created, but it can be augmented along the way with data from your own usage. The software can literally learn your style.

The same basic algorithm can handle different languages, adapt to different users and incorporate words and phrases it has never seen before, such as your name or street. The quality of its suggestions will depend mostly on the quantity and quality of data on which it is trained.

The more you use it, the more it learns the kinds of words and expressions you use. It improves its behaviour on the basis of experience, which is the definition of learning. A system of this type will probably need to be exposed to hundreds of millions of phrases, which means being trained on several million documents. That would be difficult for a human but is no challenge at all for modern hardware.

Bots in translation

The algorithms that underpin machine learning have been around for years. What's new is that we now have enough data – and computing muscle – for the techniques to gain traction.

Take language translation. In the early days of AI, linguists built translation systems based on bilingual dictionaries and codified grammar rules. But these fell short because such rules are inflexible. For example, adjectives come after the noun in French and before the noun in English – except when they don't, as in the phrase 'the light fantastic'. Translation shifted from rules that are handwritten by human experts to probabilistic guidelines that are automatically learned from real examples.

In the late 1980s IBM used machine learning to teach a computer to translate between English and French by feeding it bilingual documents produced by the Canadian parliament. Like a Rosetta Stone, the documents contained several million examples of sentences translated into both languages.

IBM's system spotted correlations between words and phrases in the two languages and reused them for fresh translation. But the results were still full of errors. They needed to be able to process more data. 'Then Google comes along and basically feeds in the entire Internet,' says Viktor Mayer-Schönberger at the Oxford Internet Institute at the University of Oxford.

On any given day, Google translates more text than all the professional human translators in the world decipher in a year. Like IBM, Google's efforts in translation started by training algorithms to cross-reference documents written in many languages. But the realization dawned that the translator's results would improve significantly if it learned how people speaking Russian, French or Korean actually conversed.

Google turned to the vast web of words it has indexed, which is fast approaching the fantastical library containing every possible sentence as imagined by Jorge Luis Borges in his story 'The Library of Babel'. Google's translator – attempting English to French, for instance – could then compare its initial attempt with every phrase written on the Internet in French. Mayer-Schönberger gives the example of choosing whether to translate the English 'light' with the French *lumière*, referring to illumination, or *léger*, for weight. Google has taught itself what the French themselves choose.

Google's translator – along with the Microsoft one Rick Rashid showed off in China, which was trained in much the same way – knows nothing about language at all, other than the relative frequency of a vast number of word sequences. And yet Google can translate with reasonable competence between 135 written languages, from Afrikaans to Zulu. Word by word, these AIs simply calculate the likelihood of what comes next. For them, it is just a matter of probabilities.

These basics are more or less intuitive. The complexity arises from the vast numbers of correlations made within enormous amounts of data. Google's self-driving cars, for example, gather almost a gigabyte of data each second to make predictions about their surroundings. And Amazon is so good at getting people to buy more because it recommends items based on billions of correlations from millions of other purchases.

The translation of Rashid's speech – working out what he had said from his voice and then translating it in an instant – shows just how powerful statistical AI can be. 'These systems don't perform miracles,' says Microsoft's Chris Bishop. 'But we're constantly surprised by how far we can get just by looking at the statistics of very large sets of data.'

'You may also like'

If you feel that this approach to intelligence is cheating because the algorithm is not really intelligent, then brace yourself. Things get worse.

The next step up in complexity from an autocomplete function is a product recommendation agent. Consider your favourite online shop. Using your previous purchases, or even just your browsing history, the agent will try to find the items in its catalogue that have the highest probability of being of interest to you. These will be computed from the analysis of a database containing millions of transactions, searches and items. Here, too, the number of parameters that need to be extracted from the training set can be staggering: Amazon has more than

200 million customers and in excess of 3 million books in its catalogue.

Matching users to products on the basis of previous transactions requires statistical analysis on a massive scale. As with autocomplete, no traditional understanding is required – it does not need psychological models of customers or literary criticism of novels. It is no wonder that some question whether these agents should be called 'intelligent' at all. But they cannot question the word 'learning': these agents do get better with experience.

Emulating behaviour

Things can get more complicated. Online retailers keep track not just of purchases but also of any user behaviour during a visit to the site. They might track information such as which items you have added to the basket but later removed, which you have rated and what you have added to your wish list. Yet more data can be extracted from a single purchase: time of day, address, method of payment and even the time it took to complete the transaction. And this, of course, is done for millions of users.

As customer behaviour tends to be rather uniform, this mass of information can be used constantly to refine the agent's performance. Some learning algorithms are designed to adapt on the fly; others are retrained offline every now and then. But they all use the multitude of signals extracted from our actions to adapt their behaviour. In this way, they constantly learn and track our preferences. It is no wonder that we sometimes end up buying a different item from the one we thought we wanted.

Intelligent agents can even propose items just to see how you respond. Extracting information in this way can be as valuable as completing a sale. Online retailers act in many ways as autonomous learning agents, constantly walking a fine line between the exploration and exploitation of their customers.

Learning something they did not know about you can be as important as selling something. To put it simply, they are curious. A similar strategy could be used by spam filters and any other software that needs to learn your preferences and predict your actions. One day soon, the appliances in your house will be interested in predicting your next action, too.

These are just the simplest examples. Using the same or similar statistical techniques, in multiple parts of a system and at various scales, computers can now learn to recognize faces, transcribe speech and translate text from one language to another. According to some online dating companies, they can even find us potential love matches. In other words, they can emulate complex human behaviours that we cannot fully model – and they do it in a way that is very different from how we would do it.

Novel situations

Machine learning is not just about analysing past behaviour. Sometimes AIs need to deal with novel situations. How do you help a new customer? To whom do you recommend a brand-new book? The trick in this case is to get machines to generalize, using information from similar customers or products.

Even a customer who has never used a service before leaves a small data trail – an email address and location, for example – to get started on. The ability to detect and exploit similarities is sometimes called pattern recognition, and its importance is not limited to 'cold-start' situations. In fact, generalization – detecting patterns and similarities – is a fundamental part of intelligent behaviour.

What do we mean when we say that two items are similar? We could describe a book by the number of pages, the language it is written in, the topic, the price, the date of publication, the author, even some index of its readability. For a customer, useful descriptors might include age, gender or location. In machine learning, these descriptors are sometimes called features or signals. We can use them to locate similar items for which we have sufficient data. The machine can thus generalize from one situation to a similar one, and make better use of its experience.

Choosing the right features is one of the critical problems in machine learning: the font used in a book might not be as useful as its price, for example. This problem becomes even more critical when we handle complex items such as images. If you compare two passport photos of yourself taken one minute apart, they will not be identical at the level of raw pixels. This is sufficient for the computer to treat them as two completely different images. We would like the computer to represent those images in a more robust way than just using pixels, so that it is not confused by small irrelevant changes in the image. Which features of an image should be used in order to recognize the same face in different photos?

This has been a surprisingly stubborn problem, made worse by the variation in lighting, position and background that can occur in natural scenes.

Programming this capability directly into a computer has proved difficult, so engineers have once more resorted to machine learning. One such method, called deep learning, is currently delivering the best results in some domains. As with the earlier examples, it involves using big data to adjust millions of parameters.

Layers of learning

One of the buzz phrases in artificial intelligence research is deep learning. It sounds exotic but is actually another form of the data-driven approach that has delivered so many successes in AI in recent years. Deep learning relies on a technology called neural networks, software circuits designed to mimic the human brain and the myriad neurons connected by synapses that realize its unparalleled computing power. In a neural net, many simple processors are wired together so that the output of one can act as the input of others. These inputs are weighted to have more or less influence, and the idea is that the network 'talks' to itself, using its outputs to alter its input weightings – in effect, learning as it goes along, just as the brain does.

In a few short years, neural networks have overtaken established technologies to become the best way to solve hard perceptual problems, from reading medical scans and recognizing faces to driving cars. Consider the task of picking out all the images of a football match from a set of photos. A programmer could write an algorithm to look for typical features like goalposts, but it is a lot of work. A neural network does that work for you, by initially finding features like the edges of objects in images, then moving on to recognizing objects and even activities. For example, a ball, a field and players are likely to indicate a football match. Each node layer looks for features at different levels of abstraction.

The gap between its output and the correct answer is fed back for it to tweak its weightings accordingly until it gets it right all – or most – of the time. When a system is trained by being given positive or negative rewards for its actions, it is known as reinforcement learning. A programmer need adjust only the number of nodes and layers to optimize how it captures relevant features in the data. However, since it is often impossible to tell exactly how a neural network does what it does, this tweaking is a matter of trial and error.

Originally based on a loose biological analogy with the human cortex, neural networks have been developed into complex mathematical objects. In their early incarnations, neural networks were not particularly useful, but with modern hardware and giant data sets they have found a new life, delivering the best performance in certain perceptual tasks – most notably in vision and speech. Deep learning is typically used as a component in larger machine-learning systems.

Under the bonnet

Consider now that these nuts and bolts of machine learning can be applied to many parts of the same system at the same time: a search engine might use them to learn how to complete your queries, best rank the answers for you, translate a document among the search results and select which ads to display. And this is just on the surface.

Unknown to users, the system will probably also be running tests to compare the performance of different methods by using them on different random subsets of users. This is known as A/B testing. Every time you use an online service, you are giving it a lot of information about the quality of the methods being tested behind the scenes. All this is on top of the revenue you generate for them by clicking on ads or buying products.

While each of these mechanisms is simple enough, their simultaneous and constant application on a vast scale results in a highly adaptive behaviour that looks intelligent to us. Google's Go-playing AI AlphaGo learned its winning strategies by studying millions of past matches and then playing against various versions of itself for millions of further matches – an impressive feat.

Nonetheless, every time we understand one of the mechanisms behind AI, we cannot help feeling a little cheated. AI systems generate adaptive and purposeful behaviour without needing the kind of self-awareness that we like to consider the mark of 'real' intelligence. Lovelace might dismiss their suggestions as unoriginal, but while the philosophers debate, the field keeps moving forward.

A new way of thinking

The data-driven approach to AI is now close to influencing every realm of life – going far beyond online shopping. A month after Rashid's speech, for example, the Netherlands Forensic Institute in The Hague employed a machine-learning system to help find a murder suspect who had evaded capture for 13 years. The software could analyse and compare large volumes of DNA samples, something that would be far too time consuming to do by hand.

The insurance and credit industries are also embracing machine learning, employing algorithms to build risk profiles of individuals. Medicine, too, uses statistical AI to sift through genetic data sets too large for humans to analyse. Systems like IBM's Watson and Google's DeepMind AI even perform medical diagnoses. Big data analysis can see things that we miss. It can even know us better than we know ourselves. But it also requires a very different way of thinking.

In the early days of AI, the notion of 'explainability' – that a system should be able to show how it reached a decision – was prized. When a rule-based, symbolic reasoning system made a choice, a human could trace its logical steps to work out why. Yet the reasoning made by a data-driven artificial mind today is a massively complex statistical analysis of an immense number of data points. It means that we have traded 'why' for simply 'what'.

Even if a skilled technician could follow the maths, it might not be meaningful. It would not reveal why it made a decision, because the decision was not arrived at by a set of rules that a human could interpret, says Microsoft's Chris Bishop. But he thinks this is an acceptable trade-off for systems that work. Early artificial minds may have been transparent, but they failed. Some have criticized this shift but Bishop and others argue that it is time to give up on expecting human explanations. 'Explainability is a social agreement,' says Nello Cristianini. 'In the past we decided it mattered. Now we've decided it doesn't matter.'

Peter Flach at the University of Bristol, UK, tries to teach his computer science students this fundamentally different way of thinking. Programming is about absolutes but machine learning is about degrees of uncertainty. He thinks we should be more sceptical. For example, when Amazon recommends a book, is that because of machine learning or because the company has books it cannot shift? And while Amazon might tell you that similar people bought the book choices it presents, what does it actually mean by 'people like you' and 'books like this'?

High stakes

The danger is that we give up asking questions. Could we get so used to choices being made for us that we stop noticing? The stakes are higher now that intelligent machines are beginning to make inscrutable decisions about mortgage applications, medical diagnoses and even whether you are guilty of a crime.

What if a medical AI decides that you will start drinking heavily in a few years' time? Would doctors be justified in withholding a transplant? It would be hard to argue your case if no one knew how the conclusion was reached. And some may trust the AI more than others. 'People are too willing to accept something that an algorithm has found out,' says Flach. 'The computer says "No". That's the issue.'

There could be an intelligent system somewhere making its mind up right now about what kind of person you are – and will be. Consider what happened to Latanya Sweeney at Harvard University. One day she was surprised to find that her Google search results were accompanied by adverts asking 'Have you ever been arrested?' The ads did not appear for white colleagues. This prompted a study showing that the machine learning behind Google's search was inadvertently racist. Deep within the chaos of correlations, names more commonly given to black people were linked to ads about arrest records.

Blunders

In recent years we have had several such blunders. In 2015 Google apologized after one of its products automatically labelled photos of two black people 'gorillas'. A year later Microsoft had to withdraw a chatbot called Tay because it had learned offensive language. In both cases, it was not a failure of the algorithm but of the training data that had been fed to it.

And 2016 also saw the first fatality linked to an autonomous car, when a driver put a Tesla on autopilot and it failed to detect a trailer on the road. The conditions were unusual, with a white obstacle against a light sky, and the computer vision system simply made a mistake. As many companies move into this market, the chances of other such incidents occurring increase.

There are also countless stories that do not end up in the news, because the AI systems are doing their work as expected. We typically have no way of knowing whether they are doing quite what we would like, however. As we entrust machines with increasingly sensitive decisions, we need to pay careful attention to the kind of data we feed to them. It is not just the technology but its deployment in our everyday life that needs better understanding.

Many people have expressed concerns about privacy in an age of big data. But Viktor Mayer-Schönberger at the Oxford Internet Institute thinks we should be more worried about the abuse of probabilistic prediction. 'There are profound ethical dilemmas,' he says.

To navigate this world, we will need to change our ideas about what AI means. The iconic intelligent systems we have built neither play chess nor plot humanity's downfall. 'They are not like HAL 9000,' says Nello Cristianini, a professor of artificial intelligence at the University of Bristol, UK. They have gone from chaperoning our time online and nudging us towards a purchase to promising to predict our behaviour before we know it ourselves. We cannot avoid them. The trick will be to accept that we cannot know why these choices were made, and to recognize the choices for what they are: recommendations, mathematical possibilities. There is no oracle behind them.

When people dreamed of making AI in our image, they may have looked forward to meeting these thinking machines as equals. The AI we have now is alien – a form of intelligence we have never encountered before.

Can we peer inside an AI's head?

A penny for 'em? Knowing what someone is thinking is crucial for understanding their behaviour. It's the same with artificial intelligences. A new technique for taking snapshots of neural networks as they crunch through a problem will help us fathom how they work, leading to AIs that work better – and that are more trustworthy.

In the last few years, deep-learning algorithms built on neural networks have driven breakthroughs in many areas of AI. The trouble is that we do not always know how they do it. A deep-learning system is a black box, says Nir Ben Zrihem at the Israel Institute of Technology in Haifa. 'If it works, great. If it doesn't, you're screwed.'

Neural networks are more than the sum of their parts. They are built from many very simple components – the artificial neurons. 'You can't point to a specific area in the network and say all of the intelligence resides there,' says Zrihem. But the complexity of the connections means that it can be impossible to retrace the steps a deep-learning algorithm took to reach a given result. In such cases, the machine acts as an oracle and its results are taken on trust.

To address this, Zrihem and his colleagues created images of deep learning in action. The technique, they say, is like an fMRI for computers, capturing an algorithm's activity as it works through a problem. The images allow the researchers to track different stages of the neural network's progress, including dead ends.

To get the images, the team set a neural network the task of playing three classic Atari 2600 games: *Breakout*, *seaQuest DSV* and *Pac-Man*. They collected 120,000 snapshots of the deep-learning algorithm as it played each of the games. They then mapped the data using a technique that allowed them to compare the same moment in repeated attempts at a game.

The results look a lot like scans of real brains (see Figure 2.1). But in this case, each dot is a snapshot of a single game at a moment in time. Different colours show how well the AI was doing at that point in the game.

With *Breakout*, for example – where the player must knock a hole through a wall of brightly coloured blocks with a paddle and a ball – the team was able to identify a clear banana-shaped region in one map showing every time the algorithm tried tunnelling through the blocks to force the ball to the top of the wall, a winning tactic that the neural network had figured out by itself. Mapping the playthroughs let the team trace how the algorithm successfully applied it in successive games.

Building the perfect game strategy is fun, but scans like these could help us hone algorithms designed to solve real problems. For example, a security algorithm might have a flaw that means it's easily fooled in certain situations, or an algorithm designed to decide whether someone gets a bank loan might be prejudiced against people of a particular race or gender. If you deploy this technology in the real world, you will want to understand how it works and where it might fail.

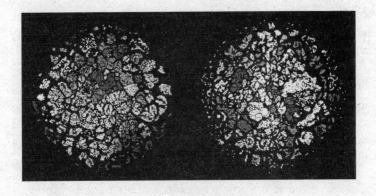

FIGURE 2.1 'Brain scans' of a neural network

The symbols strike back

There is no question that machine learning via neural networks has had unrivalled success – but it is not perfect. Training a system to do a specific task is slow and it cannot reuse what it has learned to do a different task. This problem dogs modern artificial intelligence. Computers can learn without our guidance, but the knowledge they acquire is meaningless beyond the problem they are set. They are like a child who, having learned to drink from a bottle, cannot even begin to imagine how to drink from a cup.

At Imperial College London, Murray Shanahan and colleagues are working on a way around this problem using the old, unfashionable approach that machine-learning techniques pushed to the sidelines. Shanahan's idea is to resurrect symbolic AI and combine it with modern neural networks.

Symbolic AI never took off because manually describing everything an AI needed to know quickly proved overwhelming.

Modern AI has overcome that by instead learning its own representations of the world. Yet the representations cannot be transferred to other neural nets.

Shanahan's work aims to let some knowledge transfer between tasks. The prize is AI that learns quickly and requires less data about the world. As Andrej Karpathy, a machine-learning researcher with the firm OpenAI, put it in a blogpost: 'I don't actually have to experience crashing my car into a wall a few hundred times before I slowly start avoiding how to do so.'

A higher state of mind

If we want to build a computer that performs with human-level intellect, why not make an artificial brain? Humans, after all, are our best example of intelligence and neuroscience is giving us many new insights about how we process and store information.

The human brain is a network of 100 trillion synapses that connect 100 billion neurons, most of which change their state between 10 and 100 times per second. Our brain's layout makes us good at tasks like recognizing objects in an image.

A supercomputer, on the other hand, has about 100 trillion bytes of memory and its transistors can perform operations 100 million times faster than a brain. This architecture makes a computer better for quickly handling highly defined, precise tasks.

But some tasks would benefit from more brain-like processing, even with the attendant trade-offs. For example, uncertain tasks like recognizing faces don't necessarily require highly accurate circuitry in which processing follows a precise path.

Some researchers are looking into brain-like hardware architectures to mimic the brain's low-power requirements. The brain does all of its computations on roughly 20 watts, the equivalent of a very dim light bulb. A supercomputer capable of roughly analogous computations requires 200,000 watts. Other groups are interested in learning from the brain's ability to process information and store it in the same place. For these reasons, projects are under way to build novel computer circuits inspired by the brain: more parallel than serial, more analogue than digital, slower — and consuming far less power.

Intuitive thinking

Humans persistently fail to live up to the ideal of rationality. We make common errors in our decision-making processes and are easily influenced by irrelevant details. And when we rush to a decision without reasoning through all the evidence, we call this trusting our intuition. We used to think the absence of such human quirks made computers better, but recent research in cognitive science tells us otherwise.

Humans appear to have two complementary decision-making processes, one slow, deliberate and mostly rational, the other fast, impulsive, and able to match the present situation to prior experience, enabling us to reach a quick conclusion. This second mode seems to be key to making human intelligence so effective.

While it is deliberative and sound, the rational part requires more time and energy. Say an oncoming car starts to drift into your lane. You need to act immediately: sound the horn, hit the brakes or swerve, rather than start a lengthy computation that would determine the optimal but possibly belated act. Such shortcuts are also beneficial when there is no emergency. Expend too much brain power computing the optimal solution to details like whether to wear the dark-blue or the midnight-blue shirt, and you will quickly run out of time and energy for the important decisions.

So should AI incorporate an intuitive component? Many AI systems do have two parts, one that reacts instantly to the situation and one that does more deliberative reasoning. Some robots have been built with low-level layers that are purely reactive and higher layers that inhibit these reactions and organize more goal-directed behaviour. This approach has proved to be useful for getting walking robots to negotiate rough terrain, for example.

There has been a similar push to motivate AIs to make better decisions by giving them emotions. For example, if an autonomous robot tries the same action a few times and fails repeatedly, a 'frustration' circuit would be an effective way of prompting it to explore a new path.

Creating machines that simulate emotions is a complicated undertaking. Marvin Minsky, one of the founders of AI, has argued that emotions arise not as a single thing that brains do but as an interaction involving many parts of the brain, and between the brain and the body. Emotions motivate us to choose certain decisions over others, and thinking of the parts of a computer program as if they were motivated by emotions may help pave the way for more human-like intelligence.

'Humans rarely get totally stuck because we have many different ways to deal with each situation or job,' says Minsky. 'Whenever your favourite method fails, you can usually find a different approach. For example, if you get bored with one particular job, you can try to persuade someone else to do it or get angry with those who assigned it to you. We might call such reactions emotional, but they can help us deal with the problems we face.'

Sloppy bits

Krishna Palem, a computer scientist at Rice University in Houston, Texas, is one of a handful of researchers building low-powered – brain-like – computers. His devices will not win any awards for accuracy; most of the time they cannot even add up correctly. For them, $2 + 2$ might as well be 5. But don't be fooled by the wobbly arithmetic. Palem is making machines that could represent a new dawn for computing.

Inaccuracy is not something we typically associate with computers. Since Turing laid down their ground rules in the 1930s, computers have been sticklers for precision, built on the principle of following step-by-step instructions in an exact and reproducible manner. They should not make errors.

But maybe we should let computers make mistakes: it could be the best way to unlock the next wave of smart devices and prevent high-performance computing from hitting a wall. It would allow us to run complex simulations that are beyond today's supercomputers – models that better predict climate change, help us design more efficient cars and aircraft, and reveal the secrets of galaxy formation. They may even unlock the biggest mystery of all, by letting us simulate the human brain.

Until now, we have had to accept a trade-off between performance and energy efficiency: a computer can be either fast or low-powered, but not both. This not only means that more powerful smartphones need better batteries but also that supercomputers are energy guzzlers. Next-generation 'exa-flop' machines, which are capable of 1,018 operations a second, could consume as much as 100 megawatts, the output of a small power station. So the race is on to make computers do more with less.

One way is simply to reduce the amount of time a computer spends executing code, because taking less time means that less power is used. For programmers, this means looking for ways to get the desired result more quickly. Take the classic travelling-salesman problem of finding the shortest route around a group of cities. It's notoriously tough to solve, given that the number of possible routes shoots up exponentially with the number of cities. Palem says that coders often settle for a route that they estimate to be about half as good as the best, because to do better would use up too much computer time. A more recent version of this approach is to use a machine-learning algorithm to arrive at an approximate result for a given piece of code. This rough answer – like a back-of-the-envelope calculation – can then be used each time the program runs, instead of executing the original section of code itself.

But saving energy by cutting corners in software only goes so far. To really save power, you need to change the way the hardware works. Computers can save vast amounts of energy simply by not operating all their transistors at full power all the time, but this means sacrificing accuracy. Palem's team is hobbling computers so that they get their sums wrong in an acceptable way. Take any algorithm that you think does a good job, and he will solve it inexactly with a different physical system under the hood.

On and off again

Standard computer chips use a sliver of silicon called a channel to act as a switch that can flip between on (1) and off (0). The switching is controlled by a gate that stops a current flowing through the channel until you apply a voltage. Then the gate opens like a sluice in a dam, letting current through. But this complementary metal-oxide semiconductor (CMOS) technology only works well when it has a reliable 5-volt power supply. Start to lower that and the channel becomes unstable – sometimes switching, sometimes not.

In 2003 Palem, then at the Georgia Institute of Technology in Atlanta, saw trouble coming. It was clear that the ability of the electronics industry to continue doubling the number of transistors on a chip every 18 to 24 months – a miniaturization trend known as Moore's law – was coming to an end. Miniaturization was introducing errors at the chip level. This was largely due to overheating and interference, or crosstalk, between the densely packed transistors. Power was now the critical issue. What if you could harness instabilities in some way that would also save energy?

Palem's answer was to design a probabilistic version of CMOS technology that was deliberately unstable. His team built digital circuits in which the most significant bits – those representing values that need to be accurate – get a regular 5-volt supply but the least significant bits get 1 volt. As many as half the bits representing a number can be hobbled like this.

This means that Palem's version of an adder, a common logic circuit that simply adds two numbers, doesn't work with the usual precision. 'When it adds two numbers, it gives an answer that is reasonably good but not exact,' he says. 'But it is much cheaper in terms of energy use.'

Not pixel perfect

Spread that over billions of transistors and you have a significant power saving. The trick is to choose applications for which the least significant bits do not matter too much: for example, using a large range of numbers to represent the colour of a pixel. In one experiment, Palem and his colleagues built a digital video decoder that interpreted the least significant bits in an imprecise way when converting pixel data into screen colours. They found that human viewers perceived very little loss in image quality. 'The human eye averages a lot of things out,' says Palem. 'Think about how we see illusions. The brain does a lot of work to compensate.'

Encouraged by that success, the Rice University researchers have moved on to another application involving the senses: hearing aids. Their initial tests show that inexact digital processing in a hearing aid can halve power consumption while reducing intelligibility by only 5 per cent. The results suggest that we could use such techniques to slash the power used by smartphones and personal computers, given that these are basically audio-visual devices. Many applications of AI, such as image recognition or translation, could also benefit.

Cloud atlas: improving climate predictions

Tim Palmer, a climate physicist at the University of Oxford, sees great potential in cutting computers some slack. He thinks that computers based on Palem's ideas could be the answer to what is presently an intractable problem: how to improve the accuracy of climate predictions for the next century without waiting years for a new generation of supercomputers. 'The crucial question about climate change centres on the role of clouds,' says Palmer, 'in terms of whether they amplify or dampen the effects of global warming. You can't really answer that question with any great confidence unless you can simulate cloud systems directly.' And right now, it's not clear how we do that.

Today's supercomputers lack the power to do it and their successors, expected in the next decade or so, will just be too energy-hungry. 'Based on current estimates, the amount of power needed for such a machine is going to be around 100 megawatts,' says Palmer, five to ten times what today's top supercomputers use. Assuming they don't just melt, running them could prove prohibitively costly.

Supercomputers burn so much power because they are generally optimized for computing with 64-bit-long numbers. In principle, this gives greater accuracy. But climate models involve millions of variables, simulating complex interacting factors such as winds, convection, temperatures, air pressures, ocean temperatures and salinity. The result, says Palmer, is that they have too much power-draining data to crunch. What's needed, he thinks, is for different variables to be represented in data strings of varying length, depending on their importance to the model.

The pay-offs could be huge. Today's climate models tackle Earth's atmosphere by breaking it into regions roughly 100 kilometres square and a kilometre high. Palmer thinks inexact computing would get this down to cubes a kilometre across – detailed enough to model individual clouds.

'Doing 20 calculations inexactly could be much more useful than 10 done exactly,' says Palmer. This is because, at 100-kilometre scales, the simulation is a crude reflection of reality. The computations may be accurate but the model is not. Cutting precision to get a finer-grained model would actually give you greater accuracy overall. 'It is more valuable to have an inexact answer to an exact equation than an exact answer to an inexact equation,' he says. 'With the exact equation, I'm really describing the physics of clouds.'

Degrees of accuracy

You cannot just give up on accuracy across the board, of course. The challenge is choosing what parts of a computation can be treated more crudely than others.

Researchers are attacking the problem from several different angles. Mostly, it comes down to devising ways to specify thresholds of accuracy in code so that programmers can say when and where errors are acceptable. The software then computes inexactly only in parts that have been designated safe.

Some think that inexact simulations could ultimately help us understand the brain. Supercomputers like IBM's Blue Gene are being used to model neurological functions in the Human Brain Project, for example. As we have seen, there is a huge discrepancy in power consumption between the brain and a supercomputer: while a supercomputer needs megawatts of power, a human brain runs on the power of a light bulb. What could account for this?

Tim Palmer and colleagues at the University of Sussex in Brighton, UK, are exploring whether random electrical fluctuations might provide probabilistic signals in the brain. Palmer's theory is that this is what lets it do so much with so little power. Indeed, the brain could be the perfect example of inexact computing, shaped by pressure to keep energy consumption down.

What is clear is that, to make computers better, we need to make them worse. And if approximate computing seems a shaky foundation on which to build the future of computing, it's worth remembering that computers are always dealing with the abstract. In a sense, all computing is approximate. Some computers are just more approximate than others.

Embodied intelligence

It's so obvious that it is often overlooked: we are not disembodied minds. There are good reasons to think that our intelligence is tied up with the way we sense and interact with the world. That's why a few AI researchers have always insisted that machines that think need bodies.

In January 2011 Max Versace and Heather Ames were busy with two newborns: their son Gabriel and Animat, a virtual rat. Like all babies, when Gabriel was born his brain allowed him to do only simple things like grasp, suck and see blurry images of his parents. The rest was up to him.

Animat did not start with much programming, either. But its interaction with its virtual world soon taught it how to tell colours apart and understand the space around it. Versace and Ames, both at Boston University, hope their approach will advance machine intelligence to the point that robots start to think in a more human way.

The belief that this is the right path for AI goes back decades. In the 1980s Rodney Brooks at the Massachusetts Institute of Technology had argued that it was backwards to start by programming complex abilities, when we didn't even know how to make a rudimentary intelligence that could avoid bumping into walls. Instead, he said, we should emulate nature, which has given us senses that allow us to survive independently in an unscripted world.

Brooks' idea worked. In 1989 he built Genghis, a six-legged insect-robot that was able to navigate without the help of a central control system. Its sensors responded in real time to feedback it gleaned by interacting with its environment. As the robot walked around, for example, its force inputs changed, and these changes, in turn, steered its next movements, allowing it to negotiate terrain it had not been explicitly programmed to expect.

Over the next decade, research in neurobiology, cognitive science and philosophy suggested that Brooks' ideas applied much more broadly. In the late 1990s George Lakoff, a cognitive scientist at the University of California, Berkeley, proposed that human intelligence is also inextricably linked to the way our bodies and senses interact with the environment. According to Lakoff and his supporters, our 'embodied mind' explains not only rudimentary intelligence, such as how we learn to visually recognize objects, but even complicated, abstract thought. Here, at last, was the key to building a sophisticated, human-like intelligence.

There was just one problem: embodied AI is difficult to upgrade. Improving a robot's sensor-laden body requires not just the programming of extra functions but also the painstaking disassembly and reassembly of the sensors themselves. Despite these obstacles, a few researchers found the idea too compelling to abandon. In 2009 Owen Holland at the University of Sussex in Brighton, UK, created a humanoid robot modelled on some of the principles that had led to Genghis, called Eccerobot. However, Eccerobot has exhibited no sign of intelligence. And so, even as advances in computing power and data-driven approaches have supercharged conventional AI, embodied AI limps on in ever-smaller circles.

Then Versace, Ames and their team began to realize that there was hope for embodiment after all – if you skipped the physical body. Thanks to powerful new graphics cards, video games designers can simulate anything, including a robot's body, the environment it lives in and even the complex physics underlying the interactions between the two.

The team seized on these advances to 'cheat' at embodiment. Instead of toiling over a real body, they built a virtual one, whose synthetic sensors would interact with a painstakingly rendered virtual environment. This way, they reasoned, they could reap all the advantages of embodied AI with none of the drawbacks. If it worked, they would be able to hit fast-forward on the evolution of an embodied intelligence.

Animal intelligence

Animat was born the day Versace's team hooked up its brain, which is made up of hundreds of neural models – for colour vision, motor function and anxiety, among others – all of which are faithful imitations of biology. That means they don't contain a list of explicit commands, just as Gabriel's brain did not calculate his cot dimensions to figure out where to reach for a toy.

So, like Genghis, Animat depends on the feedback it gets from its virtual body, which is equipped with the kinds of sensors found in skin and the retina, to learn and move. Unlike Genghis, however, every bit of Animat can be upgraded in the blink of an eye.

Its environment also obeys the laws of real-world physics, including gravity, giving Animat realistic sensory information. Light hits its virtual retina, for example, giving it colour vision, and correctly calibrated forces – for things like water and air pressure – are exerted against its simulated skin. Different combinations of such inputs drive Animat's reactions.

Animat's virtual world is a giant blue swimming pool surrounded by many poles, all different colours (see Figure 2.2). Like real rats, Animat hates water, courtesy of an anxious streak the researchers included in its neural models. The only way to escape the water is to find a small platform hidden somewhere below the water's surface. The test of Animat's intelligence is how quickly it learns to find that anxiety-alleviating platform.

The first experiment looked like a failure: Animat swam frantically in random patterns for an hour before the researchers gave up and ended the test. When they dropped it into the pool a second time, however, Animat's swimming pattern changed. This time, after 45 minutes of swimming in a new pattern, it stumbled on to the platform. Its anxiety levels decreased sharply when it left the water, and that reward strengthened the connections that led it there. It now knew the colour of the poles near the platform, for example, and the approximate path it took to get there.

Experiment 1

Animat swims around randomly and does not find the platform.

Experiment 2

Animat swims in a different pattern and eventually finds the platform.

Experiment 3

Animat can now find the platform faster, guided by the colours of the poles.

Experiment 4

On the fourth attempt, Animat swims directly to the white pole to find the platform.

FIGURE 2.2 The virtual Animat is programmed to have a real rat's colour vision, navigation abilities and hatred of water. The only way out of the water is to find the hidden platform by the white pole.

Sure enough, the third time it was dropped into the water, Animat spent far less time swimming before it found the platform because it looked for the right-coloured poles. On the fourth try, it didn't even hesitate: it swam directly to the platform.

Auspicious as these early experiments may be, however, the virtual world is just a rehearsal space. The real test will come when a brain trained in a virtual body is transplanted into a real one. After all, the ultimate goal is a robot that can move independently in the real world.

Mars rat

The possibilities for such machine intelligence explain why NASA came calling. A Mars rover with biological intelligence could learn to use its neural networks for vision, to balance itself and to navigate away from rough terrain, eliminating the need for constant human supervision. So the team prepared a virtual Mars for Animat, complete with craters.

Because Animat is designed to learn like its biological counterparts, some familiar questions crop up. Can Animat feel pain? After all, Animat gets negative reinforcement in the form of intense anxiety and positive reinforcement in the form of instant relief when it reaches the hidden platform.

Neither Versace nor Ames believe that Animat will become conscious, but the point might not be as silly as it sounds. Feeling could be a crucial bridge between intelligence and consciousness. Some cognitive scientists think that basic mechanisms of reinforcement, such as anxiety and relief, are exactly how human consciousness arises. The 'feel' of consciousness – the internal experience of seeing red or feeling pain – derives not from higher cognition but from simple interactions with the world.

Beyond the Turing Test

One problem with the Turing Test is that no one can quite agree what counts as a pass. Turing, writing in the 1950s, predicted that by the twenty-first century it would be possible for computers to pass the test around 30 per cent of the time. Some have interpreted this as the percentage of judges a machine has to fool, leading to headlines in 2014 claiming that a chatbot at the Royal Society in London had passed the test. Others see 50 per cent as a pass.

However, even if a chatbot managed to fool all the judges, it would not really tell us anything about its intelligence. That is because the results of the test also depend on the judges' level of technical understanding and choice of questions, which will influence their ratings.

As a result, most AI researchers long ago abandoned the Turing Test in favour of more reliable ways to put machines through their paces. In just the past couple of years, algorithms have started to match and even exceed human performance at tasks outside the realm of everyday conversation.

'I spend my time trying to get computers to understand the visual world rather than win a Turing Test, because I think it's a quicker route to intelligence,' says Erik Learned-Miller at the University of Massachusetts, Amherst. He is one of the people behind the Labeled Faces in the Wild (LFW) data set. A collection of more than 13,000 facial images and names taken from the web, it has become the de facto standard for testing facial recognition algorithms.

There have been vast improvements in this field, thanks to hardware and software advances in deep learning and neural networks. In 2014 Facebook published details of its DeepFace algorithm, which scored 97.25 per cent accuracy on the LFW data set, just short of the human average of 97.5 per cent.

'When they got that, people realized this is the way to go,' says Learned-Miller. According to him, it kicked off an arms race between tech's biggest names. In 2015 Google's FaceNet system hit 99.63 per cent – seemingly better than humans. That's not quite the case because it's hard to measure our performance accurately, says Learned-Miller. But it is fair to say that machines are now comparable to humans.

Big companies are also testing their algorithms on a data set called ImageNet, a more general collection of labelled images, and vying to win the Large Scale Visual Recognition Challenge, an annual contest associated with ImageNet. Microsoft has an algorithm that scores slightly higher than humans on this task.

What happens next?

Olga Russakovsky of Carnegie Mellon University (CMU) in Pittsburgh, Pennsylvania, who is one of the challenge's organizers, has pointed out that the algorithms only have to classify images as belonging to one of a thousand categories. That is tiny compared with what humans can achieve. To show true intelligence, machines will have to draw inferences about the wider context of an image and what might happen one second after a picture is taken, says Russakovsky. The latest generation of image recognition system are beginning to do just that.

When humans have to make decisions based on partial information, we try to infer what other people will do. Some researchers think we should look to games like poker, which involves reasoning in the face of uncertainties. This makes it a much harder game for machines than chess. Again, poker bots can now beat professional human players at Heads-Up No-Limit Texas Hold 'Em – one of the toughest forms of the game. 'I like poker a lot as a test, because it's not about trying to fake AI,' says Tuomas Sandholm, also at CMU. 'You really have to be intelligent to beat humans.'

Is there any life left in the Turing Test? Bertie Müller of the Society for the Study of Artificial Intelligence and Simulation of Behaviour, which administers the Loebner Prize, says the contest is held partly for tradition's sake. Were he alive today, Turing himself might not view it as the best test of intelligence, he says. For Müller, a better test might be to observe an AI in a variety of environments – a bit like putting a toddler in a room full of toys and studying what it does. And we are a long way from machines that can outsmart a toddler in her element.

Whose AI is top of the class?

Assessments straight out of the classroom are catching on. In 2015 an AI system called ConceptNet tackled an IQ test designed for preschool children, fielding questions like 'Why do we wear sunscreen in summer?' Its results were on a par with the average four-year-old. In 2016 a system called To-Robo passed the English section of Japan's national college entrance exam. At the Allen Institute for Artificial Intelligence in Seattle, Washington, Peter Clark and colleagues have honed an AI called Aristo by giving it New York State school science exams.

Not everyone is convinced. Ernest Davis, a computer scientist at New York University in New York City, points out that AI often struggles with what we would regard as common sense. From that viewpoint, ordinary exams might not be the best way to measure machines' progress. Instead, he suggests writing exams specifically for machines. The questions would be trivial for a human but too strange or obvious to be looked up online, such as: 'Is it possible to fold a watermelon?'

3
Anything you can do

How AI is outsmarting humans

Anything you can do, a machine can do better. That is not quite the case but it increasingly feels that way. Rapid advances in deep learning have led to computers whose abilities rival or exceed ours, in tasks ranging from playing games to recognizing what's in an image. These machines do things faster and at greater scale than even a large team of humans can match. And the more they do, the more they learn about how our world works.

In the game: AI takes on Go, poker and more

One of the most celebrated successes of machine learning came in 2016, when an algorithm called AlphaGo defeated South Korean master Lee Sedol at the game Go (see Figure 3.1). Most observers had thought such an AI was a decade away. Games have long been a benchmark for the performance of AI. When IBM's Deep Blue beat chess champion Garry Kasparov in 1997, it was hailed as the first step of an AI revolution. And this time was no different.

AlphaGo had first made headlines a few months earlier when DeepMind – an artificial intelligence company that Google bought in 2014 – announced that its AI had defeated European champion Fan Hui 5-0. That prompted DeepMind to challenge Lee, considered the game's dominant force of recent times.

FIGURE 3.1 In South Korea, Go is considered a form of expression akin to martial arts.

While the match against Fan had been played in secret, this second one took place with dozens of cameras and hundreds of reporters descending on the venue, the Four Seasons hotel in the heart of downtown Seoul. As Google's AlphaGo played Lee Sedol, the press filled two separate conference rooms – one with English commentary and one with Korean – and the match captured the public's imagination.

AlphaGo ended up beating Lee 4–1, shocking the Go community and causing a buzz around the world. But it was not the fact of the defeat so much as the way the AI went about it that was most striking. 'AlphaGo actually does have an intuition,' Google co-founder Sergey Brin told *New Scientist* hours after his firm's series-clinching third victory, which he had flown in to witness. 'It makes beautiful moves. It even creates more beautiful moves than most of us could think of.'

Korea feels the aftershock

After defeat comes resolve. Following its win, AlphaGo was the toast of Silicon Valley. But in South Korea the mood was different. There the game pulls in television contracts and corporate sponsors. Scholars study it full time in academies.

Watching Google's AlphaGo AI eviscerate Korean grandmaster Lee Sedol put the nation into shock, especially after the national hero confidently predicted that he would sweep AlphaGo aside. The actual result laid bare the power of AI.

'Last night was very gloomy and many people drank alcohol,' said Jeong Ahram, lead Go correspondent for the *Joongang Ilbo*, one of South Korea's biggest daily newspapers, speaking the morning after Lee's first loss. 'Koreans are afraid that AI will destroy human history and human culture. It's an emotional thing.'

It is perhaps the perceived beauty of AlphaGo's moves that ruffled the most feathers. 'This is a tremendous incident in the history of human evolution – that a machine can surpass intuition, creativity and communication, which have previously been considered to be the territory of human beings,' Jang Dae-Ik, a science philosopher at Seoul National University, told *The Korea Herald*.

'Before, we didn't think that artificial intelligence had creativity,' said Jeong. 'Now, we know it has creativity – and more brains, and it's smarter.' Headlines stacked up in the South Korean press: 'The "Horrifying Evolution" of Artificial Intelligence' and 'AlphaGo's Victory... Spreading Artificial Intelligence "Phobia".'

Some are optimistic that the impact of Lee's loss will spark a revolution in education and learning in South Korea. 'We're very weak at AI,' says Lee Seok-bong, a journalist for South Korean science website HelloDD.com. 'Up to this point, Korean people didn't know much about AI. But because of this match, every Korean knows about it now.'

Conquering territory

The game of Go involves two players placing black and white counters to conquer territory. It is played on a 19 by 19 board, which allows for 10^{171} possible layouts, versus roughly 10^{50} possible configurations on a standard 8 by 8 chessboard. To give you a sense of scale, it's estimated there are 10^{80} atoms in the universe. 'Go is probably the most complex game ever devised by man,' says DeepMind co-founder Demis Hassabis.

To play the game so successfully, DeepMind's AlphaGo software uses multiple neural networks. Its 'policy' network learns the game from a database of nearly 30 million moves made by expert human players, which teach the software to predict likely sequences of play. It then further improves by playing thousands of games against versions of itself. That network is then fed into a 'value' network that estimates the chance of winning given the current board position. Both networks then feed into a Monte Carlo tree search, a simulation that looks through the entire 'tree' of possible moves in a game to determine which path offers the best chance of victory. The networks serve to trim down the tree, ruling out unhelpful or long-shot moves and thus speeding up the Monte Carlo search. It's this pruning that gives AlphaGo its advantage.

AlphaGo's approach is markedly different from Deep Blue's, as it evaluates thousands fewer positions on each turn compared with the chess-playing software. In other words, AlphaGo has a better sense of which of the vast number of possible moves are promising, and spends its time focusing on those.

AlphaGo is also capable of flashes of brilliance, making moves that human players would be unlikely to consider. Take its 37th move in the second game against Lee, for example. After 2,500 years of humans playing Go, AlphaGo did something totally unexpected. It abandoned a group of pieces in one corner of the board to make a play in another, a strategy that went against anything a human player would do. The move appeared to unnerve Lee, who left the room for a few minutes and then spent a quarter of an hour thinking about how to respond. Some commentators initially thought AlphaGo had made a mistake, but the move turned out to seal its victory. Move 37 has been taken as evidence that AlphaGo is capable of what we might call intuition.

After its big win, AlphaGo continued to improve, training against top players – including defeated European champion Fan Hui. In May 2017 AlphaGo won all three games in a match against

Ke Jie, the current top-ranked player in the world. The Chinese Go Association awarded DeepMind's AI professional 9-dan status. But it is not just the machine that has honed its skills. Fan Hui has said that learning from the AI's non-human style has made him a far better player. After sparring with the AI for a few months, he jumped from 500 to 300 in the world rankings.

DeepMind is now eyeing up other challenges. 'Our hope is that one day these techniques can be extended to help us address some of society's toughest and most pressing problems, from climate modelling to complex disease analysis,' says Hassabis. That is pretty much what IBM said about Deep Blue, but cracking chess did not spark an AI revolution. What is different this time around is that AlphaGo's learning method makes it more of a generalist, which could prove key. The basic techniques it uses are much more applicable to other areas than Deep Blue's approach.

Interview: What's it like acting on behalf of an AI?

Google DeepMind's Aja Huang moved AlphaGo's pieces in the five games it played against Lee Sedol in 2016.

What does it feel like to be the physical avatar for an AI?

I feel very serious. I don't want to make mistakes, because it's the team's hard work. Also, I try very hard to respect Lee Sedol. He's a master.

You and Lee bowed towards each other before the first match, even though you're not AlphaGo.

It's a formal game, and we show respect for each other. I bow on behalf of AlphaGo.

Do AlphaGo's moves surprise you?

Oh yeah, of course. What?! Play here? Especially that shoulder hit on move 37 in Game 2. It showed up on the screen, and I was like, woah!

Does the way you place stones vary?

If AlphaGo is confident, I will play confidently. And on some moves that I also think are very good moves, I will play slightly heavier. Like, good move!

How does it seem for Lee?

I think it's a new experience to him. It's different from playing a human. The computer is cold. There is no emotion. So I think it probably makes him not so comfortable.

Do you sympathize with him?

I'm always on AlphaGo's side, but I do have sympathy. I can feel his pressure. He predicted he could crush AlphaGo 5-0, but it's so different from what he expected. But I respect him as a master.

High scores

It is not only board games that have fallen. In 2015 DeepMind revealed that it had developed an AI capable of learning to play video games just by watching them on a screen. Its AI gamer was initially trained to play 49 different video games from an Atari 2600, beating a professional human player's top score in 23 of them. The software is not told the rules of the game; instead, it uses an algorithm called a deep neural network to examine the state of the game and figure out which actions produce the highest total score.

The AI performed best on simple pinball and boxing games but it also scored highly on the arcade classic *Breakout*, which involves bouncing a ball to clear rows of blocks. It even managed to learn to tunnel through one column of bricks and bounce the ball off the back wall, a trick that seasoned *Breakout* players use. 'That was a big surprise for us,' says Hassabis. 'The strategy completely emerged from the underlying system.'

Watching an Atari game is the equivalent of processing about 2 million pixels of data a second. This suggests that Google is interested in using its AI to analyse its own large data sets. Since the AI learns by watching the screen rather than being fed data from the game code, one possibility is that the AI could be used to analyse image and video data.

What game should AI take on next?

Here is what the experts say:

Diplomacy

Mark Bishop at Goldsmiths, University of London, suggests the strategy board game *Diplomacy*, in which players pose as European powers competing for land and resources. Bishop says that AlphaGo 'doesn't know the meaning of any of the symbols it is so adroitly manipulating: it doesn't even know that it is playing Go.' *Diplomacy* embodies many of the obstacles between current and true AI. 'Interestingly, it is a game that in theory a computer could play very well, as moves are communicated in writing,' says Bishop. But it would have to first pass the Turing Test – humans could team up against the AI if they worked out which player it was.

StarCraft

In Go there might be about 300 possible moves at any time. In *StarCraft*, a strategy video game with hundreds of

---

pieces, there might be 10^{300}. 'You can't even examine all possible moves in the current state, let alone all possible future move sequences,' says Stuart Russell at the University of California, Berkeley. Instead, the AI would have to consider its actions and goals on a higher level and then work out a plan to get there – requiring reasoning methods applicable to a wider range of real problems.

Dungeons & Dragons

Julie Carpenter, at the California Polytechnic State University in San Luis Obispo, says, 'AlphaGo is not trying to prove or disprove a humanlike sense of reality or believability, but instead is purely goal-centred – to win the game.' She says it would be interesting to throw AI at something like a role-playing game. There, the machine's goals would not be as obvious. It would need to rely on skills like social communication and higher-level situational awareness in order to succeed.

Cheating

Human players can read their opponent's faces and body language for clues about what to do next. They can also get ahead by using deceptive tactics, like misdirection. Could a robotic hustler ever successfully spot these false behaviours – or even cheat without being detected? Ronald Arkin of the Georgia Institute of Technology in Atlanta says, 'These twists on gaming go beyond the largely mathematical challenges that are now being breached by current AI.'

The real world

Murray Shanahan at Imperial College London says, 'I'm not particularly interested in seeing AI pitted against other games. That's useful for testing an algorithm or new learning methods, but the true frontier is the real world. When machine learning is as good at understanding the everyday world as it is at Go, we'll be well on the way to human-level artificial general intelligence.'

The chips are down

In January 2017 computers claimed another gaming victory over humans, by winning a 20-day poker tournament. The AI, called Libratus, took on four of the world's best Heads-Up No-Limit Texas Hold 'Em poker players at a Pennsylvania casino. After 120,000 hands, Libratus won, with a lead of over $1.7 million in chips.

A poker-proficient AI is remarkable because poker is a game of 'imperfect information': players don't know what cards their opponents have, so never have a full view of the state of play. This means that the AI has to take into account how its opponent is playing and rework its approach so it doesn't give away when it has a good hand or is bluffing.

The win is another important milestone for AI. Libratus's algorithms are not specific to poker, or even just to games. The AI was not taught any strategies and instead had to work out its own way to play, based on the information given – in this case, the rules of poker. This means that Libratus could be applied to any situation that requires a response based on imperfect information.

The everyday world is full of imperfect information. The researchers from Carnegie Mellon University who built the AI think that it has applications in cybersecurity, negotiations, military settings, auctions and more. They have also looked at how AI can bolster the fight against infections, by viewing treatment plans as game strategies.

Libratus has three main parts. The first computes a big list of strategies the AI could use when play begins. At the outset of the tournament, Libratus had spent the equivalent of 15 million hours of computation honing its strategies. The second part, called the 'endgame solver', took into account 'mistakes' the AI's opponents made – instances where they left themselves open to exploitation – to predict the result of each hand.

The final part of the AI looked for its own strategic weaknesses so it could change how it played before the next session. This sought to identify things its opponents were exploiting, such as a giveaway 'tell' that another player had noticed. This was important because, in the last tournament, the human players were able to work out how the AI played when it had different cards and to change the way they bet accordingly.

Jason Les, one of the professional players in the tournament, described Libratus as 'insanely good' and noted that its strategy seemed to improve as time went on, making it harder and harder to beat every day.

Fast and furious

In 2016 we learned that Facebook's AI software can probably map more in a week than humanity has mapped over our entire history.

The social network announced that its AI system took two weeks to build a map that covers 4 per cent of our planet. That's 14 per cent of the Earth's land surface, with 21.6 million square kilometres of photographs taken from space, digested and traced into a digital representation of the roads, buildings and settlements they show. And

Facebook says it can do it better and faster, potentially mapping the entire Earth in less than a week. Facebook's goal is to build maps to help the social network plan how to deliver the Internet to people who are currently offline.

This is one of the starkest examples so far of the most important phenomenon in technology – computers doing human work really quickly. It is going to change the way we work for good, and will have massive implications for how we acquire knowledge, cooperate on large projects and even understand the world.

The model was able to map 20 different countries after being trained on just 8,000 human-labelled satellite photos from a single nation. The company later improved the process to the point where it could do the same mapping in a few hours. Assuming it had the photographs, it could map the Earth in about six days. Mapping on the scale that its AI system has demonstrated would take decades for a human team of any size – and it is more data than people or their organizations are built to handle.

Facebook's map-making AI is just one of probably thousands of narrow AIs – those trained to focus on a single task – churning through human tasks around the planet right now, faster and on larger scales than we ever could. The CERN particle physics laboratory near Geneva, Switzerland, is using deep learning to find patterns in the mass of its collision data; pharmaceutical companies are using it to find new drug ideas in data sets that no human could plumb. Nvidia's Alison Lowndes, who helps organizations build deep-learning systems, says she now works with everyone: governments, doctors, researchers, parents, retailers and even, mysteriously, meatpackers.

What is exciting is that all neural networks can scale like Facebook's mapping AI. Do you have a narrow AI that can spot the signs of cancer in a scan? Good: if you have the data, you can now search for cancer in every human on Earth in a few hours. Do you have an AI that knows how to spot a crash in the markets? Great: it can watch all 20 of the world's major stock exchanges at the same time, as well as the share prices of individual companies.

The real power of narrow AI is not in what it can do, because its performance is almost never as good as that of a human would be. The maps that Facebook's AI produces are nowhere near as good as those that come out of a company such as custom map developer Mapbox. But the smart systems being built in labs at Google, Facebook and Microsoft are powerful because they run on computers. What the future of human work looks like will be determined by whether it is better to do an average-quality job 50 million times a second or a human-quality job once every few minutes.

Learning to see and hear

Cameras are now everywhere – in our phones, in our homes and in most public spaces – and the world is increasingly being monitored by software. An AI that can recognize what is in a photo can help to categorize the hundreds of billions of images that we now take and upload to social media, for example. This will help us both find pictures of things we would like to see and police the vast number of images for illegal or offensive content, something that is no longer possible to do by hand. Image recognition also lets machines better understand the human world and helps them learn how to move about in it.

There is still some way to go, especially when monitoring the unpredictable real world. But in some tests AI can already identify what is in a picture – including individual faces – more accurately than we can. Still, the power of machines that can listen and watch is not that they can do this better than human eyes or ears. Their power, like all applications of computation, lies in speed, scale and the relative cheapness of processing. Most large tech companies are developing neural networks for understanding speech, opening up data sets that were previously difficult, or impossible, to search.

How the world works

Picking out the objects in an image is one thing; understanding the wider context of a scene is a lot more difficult. One way to get machines to understand the world better is to train them to predict the future. For example, researchers at Facebook are working on AI that can guess what is going to happen next when it looks at images. It can generate a few frames of video that play forward from a moment of time into a possible future.

They are not the only ones working on this technique. Teaching AI to anticipate how a situation is likely to play out can help it comprehend the present. 'Any robot that operates in our world needs to have some basic ability to predict the future,' says Carl Vondrick at the Massachusetts Institute of Technology. 'If you're about to sit down, you don't want a robot to pull the chair out from underneath you.'

Vondrick and his colleagues have trained an AI on 2 million videos from image-sharing site Flickr, featuring scenes such as beaches, golf courses, train stations and babies in hospital. Once trained, the AI can guess what happens next when shown a single image. Given a picture of a train station, it might generate a video of the train pulling away from the platform. Or an image of a beach could inspire it to animate the motion of lapping waves.

The videos can seem a bit wonky to a human eye and the AI still has much to learn. It does not realize that a train leaving a station should also eventually leave the frame, for example. This is because it has no prior knowledge about the rules of the world – what we would call common sense. The 2 million videos – about two years of footage – are all the data it has to go on to understand how everything works. That's not that much in comparison to a ten-year-old child – or how much humans have seen over many millions of years of evolution.

The team is working on generating longer videos, in which the AI will project its imagination even further ahead. It may never be able to predict exactly what will happen but it could show us alternatives. 'I think we can develop systems that eventually hallucinate these reasonable, plausible futures,' says Vondrick.

Getting a grip

One activity where machines still fall short is interacting with the physical world. While DeepMind was prepping for their big Go game, another Google team was working on a more mundane win. In a video released in 2016, robotic claws dip and grab at household objects like scissors or sponges. They repeat the task hundreds of thousands of times, teaching themselves rudimentary hand–eye coordination. Through trial and error, the robots gradually get better at grasping until they can reach for an item and pick it up in one fluid motion.

In the same week, Facebook revealed how one of its AIs was teaching itself about the world by watching videos of wooden-block towers falling. The aim was to let it acquire intuition about physical objects in much the way human infants do, rather than making judgements based on prewritten rules.

Getting machines to handle the real world with the intuition of a child is a big challenge for AI researchers. Mastering a complex game is impressive, but it is the AIs playing with kids' toys that we should be watching. Despite Go's complexity, the challenges in the game are defined by clear rules. The real world rarely affords such luxuries.

'Frankly, my five-year-old is a lot more intelligent than AlphaGo,' says Oren Etzioni, CEO of the Allen Institute for Artificial Intelligence in Seattle, Washington. 'Any human child is substantially more sophisticated, more flexible, more able to deal with novel situations, and more able to employ common sense.'

Nevertheless, the robo-claw experiment shows that the machine-learning techniques used to master Go can also teach machines hand–eye coordination. People are trying to make AIs a little more like us – improving their dexterity through feedback from their successes and mistakes. Over the course of two months, the robo-claw team filmed 14 robotic manipulators as they tried to pick up objects. These 800,000-plus 'grasp attempts' were then fed back into a neural network.

With the updated algorithm now driving the robot's choices, the researchers put their machines to the test. They filled bins with random objects, including some that would be difficult to pick up for the two-fingered grippers – Post-it notes, heavy staplers and things that were soft or small.

Overall, the robots managed to grasp something more than 80 per cent of the time. And they developed what the team described as 'unconventional and non-obvious grasping strategies' – learning how to size objects up and treat them accordingly. For example, a robot would generally grab a hard object by putting a finger on either side of it. But with soft objects such as paper tissues, it would put one finger on the side and another in the middle.

The Facebook team took a similar approach. They trained algorithms on 180,000 computer simulations of coloured blocks stacked in random configurations, as well as videos of real wooden-block towers, filmed as they fell or stayed in place. In the end, the best neural networks accurately predicted the fall of the simulated blocks 89 per cent of the time. The AI fared less well with real blocks, with the best system getting it right only 69 per cent of the time. That was better than human guesses on what would happen to virtual blocks, and the same as humans for predicting the fall of real blocks.

Studies like these start to move away from supervised learning, a standard approach to training machines that involves giving them the right answers. Instead, learning becomes the algorithm's responsibility. It takes a guess, finds out if it succeeded, then tries again. AlphaGo also trained in part through such a trial-and-error approach, helping it to make winning moves that perplex human players.

Another skill that AIs will have to master to rival a child is doing not just one task well, but many tasks. Such intelligence is likely to be decades away, says Etzioni. 'Human fluidity, the ability to go from one task to another, is still nowhere to be found.'

Hours of mundane video

Researchers at the University of Pennsylvania in Philadelphia are teaching a neural network called EgoNet to see the world through their eyes, by feeding it hours and hours of mundane video footage shot with GoPro cameras attached to people's heads.

Volunteers had to annotate videos of their day-to-day lives frame by frame, to show where their attention was focused in each scene. They then fed the footage into a computer and asked EgoNet over and over again to tell them what they were doing. That data helped train it to make predictions, picking out things that a person was about to touch or look at more closely. You are more likely to pick up a coffee mug if its handle is pointed towards you, for example. Similarly, someone wanting to use a computer will approach its keyboard first.

The team tested EgoNet on footage that included people cooking, children playing and a dog running in a park. It still has some way to go before it can rival a human, but the researchers hope a version of the system might be useful in health care, perhaps helping doctors diagnose unusual behaviour patterns in children.

In another project called Augur, researchers at Stanford University in California have also tried to get computers to understand what's happening in first-person video. But instead of learning from annotated footage, Augur was brought up on a very different data set: 1.8 billion words of fiction taken from Wattpad, an online writing community.

Events to chew over

Fiction is a great resource for making predictions about human behaviour because it describes the breadth of human life.

Stories also tend to have a narrative structure that provides a logical sequence of events for a computer to chew over.

When Augur identifies an object in a scene, it mines what it has read to guess what a person might do with it. If it spots a plate, for example, it infers that someone is probably planning to eat, cook or wash up. If you wake up and look at your alarm clock, Augur should guess that you are about to get out of bed.

One drawback of relying on fiction is that it has given Augur a dramatic bent. If a phone rings, it thinks you are about to start swearing and will throw the phone against a wall. Tweaking the system using more mundane scenarios will help teach Augur that not everyone lives inside a soap opera. The researchers think such a system could help people by screening calls when it recognizes that they are busy, or reminding the wearer of their shopping budget when it catches them eyeing up expensive goods.

Researchers at Facebook have also trained their AI on fiction. One of their data sets includes the text from dozens of classic children's books, such as *The Jungle Book*, *Peter Pan*, *Little Women*, *A Christmas Carol* and *Alice's Adventures in Wonderland*. The AI is then asked to fill in the blanks in sentences describing events from the stories to test its reading comprehension.

The Facebook researchers argue that being able to answer such questions shows that an AI can make decisions by drawing on a situation's wider context, a crucial skill in representing and remembering complex pieces of information. Similar thinking drove another Facebook-devised intelligence test that involves answering basic questions about the relationships between objects in short stories.

Interview: Can we give computers common sense?

Facebook has several artificial intelligence projects on the go. Yann Lecun, a professor of computer science at New York University and Facebook's director of AI, is building artificial neural networks that have a sophisticated understanding of images and text — what's in a picture or story, how it all comes together, and what might come next. In this 2015 New Scientist *interview he reveals what this technology can do.*

What are the big challenges ahead for you?

The big challenge is unsupervised learning: the ability of machines to acquire common sense by just observing the world. And we don't have the algorithms for this yet.

Why should AI researchers be concerned about common sense and unsupervised learning?

Because that's the type of learning that humans and animals mostly do. Almost all of our learning is unsupervised. We learn about how the world works by observing it and living in it without other people telling us the name of everything. So how do we get machines to learn in an unsupervised way like animals and humans?

Facebook has a system that can answer simple questions about what's happening in a picture. Is that trained by annotations made by humans?

It's a combination of human annotation and artificially generated questions and answers. The images already have either lists of objects they contain or descriptions of themselves. From those lists or descriptions, we can generate questions and answers about the objects that are in the picture and then train a system to use the answer when you ask the question. That's pretty much how it's trained.

Are there certain types of questions your AI system struggles with?

Yes. If you ask things that are conceptual then it's not going to be able to do a good job. It is trained on certain types of questions, like the presence or absence of objects or the relationship between objects, but there are a lot of things it cannot do. So it's not a perfect system.

Is this system something that could be used for Facebook or Instagram to automatically caption pictures?

Captioning uses a slightly different method, but it's similar. Of course, this is very useful for the visually impaired who use Facebook. Or, say you're driving around and someone sends you a picture and you don't want to look at your phone, you could ask, 'What's in the picture?'

Are there problems that you think deep learning or the image-sensing convolutional neural nets you use can't solve?

There are things we cannot do today, but who knows? For example, if you had asked me ten years ago, 'Should we use convolutional nets or deep learning for face recognition?' I would have said there's no way it's going to work. And it actually works really well.

Why did you think that neural nets weren't capable of this?

At that time, neural nets were really good at recognizing general categories. So here's a car: it doesn't matter what car it is or what position it is in. Or there's a chair: there are lots of different possible chairs and those networks are good at extracting the 'chair-ness' or the 'car-ness' independently of the particular instance and the pose.

But for things like recognizing species of bird or breeds of dog, or plants or faces, you need fine-grained recognition where you might have thousands or millions of categories, and the differences between the different categories are minute. I had thought deep learning was not the best approach for this – that something else would work better. I was wrong. I underestimated the power of my own technique. There are a lot of things that now I might think are difficult, but that, once we scale up, are going to work.

Facebook did an experiment in which engineers gave a computer a passage from *Lord of the Rings* and then asked it to answer questions about the story. Is this an example of Facebook's new intelligence test for machines?

It's a follow-up of that work, using the same techniques that underlie it. The group working on this has come up with a series of questions that a machine should be able to answer. Here is a story; answer questions about this story. Some of them are just a simple fact. If I say, 'Ari picks up his phone' and then asked the question 'Where is Ari's phone?' the system should say that it's in Ari's hands.

But what about a whole story, where people move around? I can ask, 'Are those two people in the same place?' and you have to know what the physical world looks like if you want to be able to answer that question. If you want to be able to answer questions like 'How many people are in the room now?' you have to remember how many people came into this room from all the previous sentences. To answer those questions, you require reasoning.

Do we need to teach machines common sense before we can get them to predict the future?

No, we can do this at the same time. If we can train a system for prediction, it can essentially infer the structure of the world it's looking at by doing this prediction. A particular embodiment of this that's cool is a thing called Eyescream. It's a neural net that you feed random numbers and it produces natural-looking images at the other end. You can tell it to draw an airplane or a church tower and, for things that it's been trained on, it can generate images that look sort of convincing.

So that's a piece of the puzzle, to be able to generate images – because if you want to predict what happens next in videos, you must first have a model that can generate images.

What kind of things could a model predict?

If you show a video to a system and ask, 'What's the next frame in the video going to look like?' it's not that complicated. There are several things that can happen, but moving objects are probably going to keep moving in the same direction. But if you ask what the video will look like a second from now, a lot of things can happen.

And what if you're watching a Hitchcock movie and I ask, '15 minutes from now, what is it going to look like in the movie?' You have to figure out who the murderer is. Solving this problem completely will require knowing everything about the world and human nature. That's what's interesting about it.

> **Five years from now, how will deep learning have changed our lives?**
>
> One of the things we're exploring is the idea of the personal butler, the digital butler. There isn't really a name for this, but at Facebook it's called Project M. A digital butler is the long-term sci-fi version of Facebook's virtual assistant M – like in the movie *Her*.

Loud and clear

Machines are not only learning to see but they are listening, too. Voice recognition has advanced rapidly over the last few years. We are now close to taking it for granted that we can ask our phone to search for something online or set a reminder just by talking to it. And devices like Amazon's Echo and Google Home can be controlled entirely by voice. The security industry is also investing in smart intruder alarms that can tell the difference between a smashed window and a dropped wine glass, for example.

How did voice recognition suddenly go everywhere? It's the usual story. The technology's recent leaps are thanks to machine learning and the vast piles of data available to train it. 'There have been more improvements in speech recognition over the past three years than there have been over the past 30 years combined,' says Tim Tuttle, the CEO of Expect Labs, a start-up in San Francisco that builds smart voice interfaces.

There are still a few hills to climb. Accents and noisy backgrounds sometimes still trip up the technology – as do children, who have higher voices and are more likely to break the predictable rules of grammar. But the potential for powerful voice recognition systems is huge. People with disabilities could operate machines with greater ease, while those with busy or hands–on occupations can call up a digital assistant – similar to how physicians use voice recognition to dictate health records.

The dream that many companies are now working on is a system that not only understands what we are saying but also divines and anticipates our needs, like a personal assistant. For that to work, the system would have to understand complex queries with ambiguous or imprecise words, or get better at telling people what about their request it didn't understand. It would also need to remember previous conversations – if I search for a plane ticket to Atlanta in September, for example, and then say, 'I'd like a hotel, too,' the system should figure out when and where I want one rather than making me state that information again.

A machine that can cope with the imprecision and ambiguities in everyday speech is still some years away. Even figuring out what the pronoun refers to in a sentence like 'The trophy would not fit in the brown suitcase because it was too big' is an enormous challenge.

Code word

Plug a machine-learning system into prison phone lines and you can find out secrets a human monitor would never notice. Every call into or out of US prisons is recorded. It can be important to know what's being said, because some inmates use phones to conduct illegal business on the outside. But the recordings generate huge quantities of audio that are prohibitively expensive to monitor with human ears.

To help, one jail in the Midwest used a machine-learning system developed by London firm Intelligent Voice to listen in on the thousands of hours of recordings generated every month. The software saw the phrase 'three-way' cropping up again and again in the calls – it was one of the most common non-trivial words or phrases used. At first, prison officials were surprised by the overwhelming popularity of what they thought was a sexual reference.

Then they worked out that it was code. Prisoners are allowed to call only a few previously agreed numbers. So if an inmate wanted to speak to someone on a number not on the list, they would call their friends or parents and ask for a 'three-way' with the person they really wanted to talk to – code for dialling a third party into the call. No one running the phone surveillance at the prison spotted the code until the software started churning through the recordings.

This story illustrates the speed and scale of analysis that machine-learning algorithms are bringing to the world. Intelligent Voice originally developed the software for use by UK banks, which must record their calls to comply with industry regulations. As with prisons, this generates a vast amount of audio data that is hard to search through. The company trained its AI on waveforms of human voices – its pattern of spikes and troughs – rather than the audio recording directly. Training the system on this visual representation let it benefit from powerful techniques designed for image classification.

Look who's talking

As well as getting better at understanding speech, machines are also homing in on individual speakers. The latest versions of Apple's iPhone operating system learn what your voice sounds like and can identify you when you speak to Siri, ignoring other voices that try to butt in.

Siri, the intelligent personal assistant, is not the only one who knows your voice. As learning software improves, voice-identification systems have started to creep into everyday life, from smartphones to police stations to bank call centres. More are probably on the way. Researchers at Google have unveiled an artificial neural network that could verify the identity of a speaker saying 'OK Google' with an error rate of 2 per cent. Your voice is a physiological phenomenon shaped by your physical characteristics and the languages you speak. It is different from anyone else's, even family members – like your fingerprints or your DNA. Machine-learning techniques can tease apart the tiny differences.

Recognizing individual voices is different from understanding what they are saying. The recognition software has been fuelled by massive sets of vocal data built into a huge model of how people speak. This allows measurements of how much a person's voice deviates from that of the overall population, which is the key to verifying a person's identity. The software can be thrown off by changes to someone's voice due to sickness or stress, however.

The technology is already being used in criminal investigations. In 2014, when journalist James Foley was beheaded, apparently by ISIS, police used it to compare the killer's voice with that of a list of possible suspects. And the banks J. P. Morgan and Wells Fargo have reportedly started using voice biometrics to figure out whether people calling their helplines are scam artists.

Researchers are now figuring out how to build profiles of a stranger from audio recordings. A voiceprint gives insight into the speaker's height and weight, their demographic background, and even what their environment is like. Working with doctors, it might also be possible to detect a person's likely diseases or psychological state through voice analysis.

Rhetorical devices: AI learns to argue

In Douglas Adams' novel *Dirk Gently's Holistic Detective Agency*, a computer program called Reason can retroactively justify any decision, providing an incontrovertible argument that whatever was decided was the right thing to do. The software proves so successful that the Pentagon buys it lock, stock and barrel, shortly before a dramatic increase in public approval of military spending.

We're not quite there yet. Machines have beaten us at logical games like Go and games of bluff and chance like poker. But no computer has ever come close to beating humans where it counts: in an argument.

The first wave of artificial intelligence, able to crunch huge amounts of information and spot interconnections ever more efficiently, gave us search engines such as Google's. A machine able to formulate an argument – not just searching information, but also synthesizing it into more or less reasoned conclusions – would take the search engine to the next level. Such a 'research engine' could aid decision-making in arenas from law to medicine to politics. And with an array of ongoing projects looking to build an argumentative streak into AI, it seems only a matter of time before we'll be testing our mettle against silicon here, too.

Arguing is something humans are peculiarly good at. From polite disagreements over the dinner table to vein-popping run-ins over a parking space or presidential politics, exchanging contrary views is what we do. There are few conversations in which not a single argument is exchanged. Arguing is a human universal. As the world our ancestors lived in grew in complexity, individuals who questioned the truth of one another's claims would have had a powerful evolutionary advantage. Argumentativeness could even be the fount of all rational thought: our ability to ponder a situation's pros and cons may have originated in rehearsals for these showdowns.

And that's a fact

The social roots of human argumentation make it tough for an AI. Even when IBM's Watson wiped the floor with two human champions of the quiz show *Jeopardy!* it was only showing off an unimaginative ability to answer factual questions.

In the messy real world, such techniques can only get you so far. 'A lot of the questions we encounter in life are not factual,' says Noam Slonim at the IBM Haifa Research Lab in Israel. 'Questions where there is no one clear answer.'

Since the *Jeopardy!* success, Slonim has been collaborating with the Watson team to see whether a machine could graduate from facts to arguments. If you were to ask it, for example, whether violent video games should be sold to children, it would synthesize facts into arguments for and against the idea, instead of merely presenting you with links to other people's opinions.

Users would still have to decide which arguments to trust, just as we decide which links to trust from a search engine. But in a world where we are often swamped with information, an argument engine could save lawyers the trouble of trawling through vast archives in search of legal precedents, when a simple press of a button would produce an ironclad summary. Doctors could plug in symptoms and get robust recommendations from case histories on file. Companies might use machines to create arguments for buying their wares. Politicians could secretly test the strength of their manifestos. We might even consider consulting an argument engine before we vote.

All of this means that Slonim is no longer working alone. He now has a team of more than 40 people, and other research groups are springing up around the globe.

Appeals to reason

The first question Slonim's team had to tackle was what an argument is, logically speaking. A rough answer might be that it is a claim backed up by evidence. But then the word 'claim' itself could do with a definition. Producing a foolproof spotter's guide for an AI is surprisingly tough.

To train Watson, Slonim and his team turned to Wikipedia, surmising that the online encyclopaedia's entries would be a rich source of claim and counterclaim. It turned out to be a gargantuan task – less like looking for a needle in a haystack and more like searching for specific pieces of hay. 'In Wikipedia you have something like 500 million sentences,' says Slonim. 'And a claim is not a sentence. A claim is usually hiding within a single sentence.'

The work has begun to identify key features that set claims apart from generic statements. Claims, for example, are more likely to mention specific times and places, and to include sentiment words such as 'exceptional' or 'strong'. Later, the team hopes to shift their focus to flagging up evidence that supports the claim, as well as teaching the system to distinguish between anecdotal data and expert testimony – and learn how much weight to give different forms of evidence.

That is all very well when we want a logical, dispassionate assessment of the facts. But it is rarely facts alone that win people over – logical discussions tend to be trumped by how we feel about something.

Once more with feeling

Any artificial intelligence that aspires to rise beyond a mere fact-driven research machine to become a fully fledged 'argument machine' – one that doesn't just argue but argues with human guile – must master these elements of argument, too. But why would we want one?

Francesca Toni, an AI researcher at Imperial College London, says that argumentation is the way to resolve conflict. Machines capable of this could help us evaluate conflicts better and more easily, avoiding mistakes. Chris Reed, an AI researcher at the University of Dundee, UK, thinks that is a little utopian. But he agrees that argument machines could help raise the level of public discussion.

Over the past few years, Reed and his team have been on a mission to seek out good arguments, then dissect and rework them into a form that can be used to train an AI to argue as we do. This quest has led them to some unexpected places. The boisterous debates in the UK parliament, for example, are not good reference material: too much performative swagger and too many procedural interjections and references to previous debates. 'There is much less quality argument there than you would either expect or hope,' says Reed. Some online forums, on the other hand, contain surprisingly well-structured arguments, despite users being inclined to wear their heart on their sleeve.

Reed's favourite source is the BBC radio show *Moral Maze*, in which panellists debate the ethics of an issue of the day. Its quasi-legal cut and thrust, laced with appeals to our emotions, is just the thing from which to build a general framework for the essence of human argument. By analysing and providing a classification of the sorts of arguments we use and how they relate to one another, Reed and his team aim to produce a tool that can then be used to train an AI.

In July 2012 they performed their first real-time argument analysis, of a *Moral Maze* episode on the ins and outs of British colonial rule in Kenya. Claim and counterclaim, and the connections between them, were represented on a giant touchscreen in a form ready to be fed into an AI. His team has since repeated the exercise many times, dissecting episodes of *Moral Maze* and other broadcast and print sources, plus some online forum postings, and turning them into a public databank of argument maps.

Changing minds

Reed and his team have also started collaborating with IBM in an attempt to build Watson's familiarity with webs of human reasoning. Meanwhile, a project started by Ivan Habernal and Iryna Gurevych at the Technical University of Darmstadt in Germany aims to go a step further, analysing not just what sort of arguments we use but which ones are the most effective. In 2016 they asked nearly 4,000 people to say which of two arguments, each making the same case in different ways, they found the more convincing – and to explain why. Over 80,000 responses later, they now have a database that can be used to teach computational systems to rank the arguments they process, and so argue more convincingly. 'To me the goal is to change somebody's mind, to persuade them,' says Habernal.

Will we swallow it? A full-blown argument machine sounds as implausible as Douglas Adams' tongue-in-cheek Reason software. It seems hard to believe that anyone would trust a machine to tell them, say, how to vote, or suggest what they should think about a certain issue. Then again, if anyone had said just over two decades ago that we would trust an AI to serve up and rank sources of information for us online, few people would have believed that either.

4

Matters of life and death

Driverless cars, AI doctors and killer robots

Machines that can understand the world around them have the potential to reshape it. This opens up exciting new possibilities, such as completely rethinking how transport works or how quickly we can treat diseases. But it also raises the spectre of new types of weapon. Driverless cars, smart medical devices and so-called killer robots all have the potential to save lives - but they could also cause great harm. Can the ethical debates keep up with the technology?

Driverless cars

Until now, driving has been a task best left to humans precisely because it involves so many variables. Is the approaching car going at 60 or 70 kilometres per hour? Could there be another vehicle out of sight around the corner? If I attempt to pass the car ahead, will its driver speed up?

For AI, the actual driving has been the easy part, at least on main roads. In 1994 two driverless Mercedes-Benz cars fitted with cameras and onboard AI drove 1,000 kilometres on roads around Paris. However, most driving takes place in cities, and that's where it becomes tough for AI, which until recently was unable to negotiate the unwritten rules of city traffic. For example, when Google's researchers programmed an autonomous vehicle to faithfully give way at an intersection as specified in the driver's manual, they found that the self-driving car would often never get a chance to go. So they changed the car's behaviour to make it inch forward after it had waited a while, signalling its intent to move ahead.

Another major source of uncertainty for a self-driving car is locating itself in space. It can't rely solely on GPS, which can be off by several metres, so the AI compensates by simultaneously keeping track of feedback from cameras, radar and a range-finding laser, cross-checked against GPS data. An average of these imperfect locations provides a highly accurate measurement.

AI is not restricted to driving. In recent-model cars, an AI program automatically adjusts the fuel flow to make it more efficient and the brakes to make them more effective. The most advanced cars of today boast systems to help you cruise down a motorway, creep along in bumper-to-bumper traffic or detect hazards in poor light using thermal imaging. Some cars can even do the dreaded parallel parking or prevent you from rear-ending the vehicle in front. They are still unable to take your non-driving grandparent to bingo, pick the children up from school or let you work peacefully in the back – but with the world market for driverless tech projected to grow 16 per cent per year over the next decade, that is probably only a short ride away.

Into fifth

Not all so-called autonomous vehicles are created equal. SAE International, a globally recognized standards organization for the transport industry, has developed a widely used scale that measures different degrees of autonomy. Here is how the scale breaks down:

Level 0: No autonomous features; may have automatic gear shift. *Most cars currently on the road.*

Level 1: Some autonomous features, e.g. automatic braking, cruise control. *Many newer car models.*

Level 2: Automated steering, braking and acceleration, but requires human oversight. *Tesla Model S, Mercedes-Benz 2017 E Class, Volvo S90.*

Level 3: Car can monitor its environment and drive autonomously, but may request human intervention at any time. *Audi A8 (2018), Nissan ProPILOT 2.0 (2020), Kia DRIVE WISE (2020).*

Level 4: Car can drive independently but may request human intervention in unusual conditions, e.g. extreme weather. *Volvo (2017), Tesla (2018), Ford (2021), BMW iNext (2021).*

Level 5: Car can drive independently in all conditions. *Driverless pods like those being rolled out in London's Gateway project.*

FIGURE 4.1 The driverless pods being rolled out in London do not look like other cars.

Finding the way

Fleets of experimental self-driving cars have already logged hundreds of thousands of kilometres on highways and busy city streets, with no human intervention. Now the training wheels are coming off. Several cities around the world are rolling out driverless vehicles as part of their public transport network.

London will be one of the first. Thanks to the Gateway project, you will soon be able to jump into an autonomous pod in Greenwich and be ferried to your destination along public roads (see Figure 4.1). After several years of hype, this will be the first chance most people will have to experience driverless cars for themselves – not just the passengers but also those sharing the roads with them.

These small pilot projects in urban areas are the beginning of a revolution in transport. In the UK, Greenwich, Milton Keynes, Coventry and Bristol will lead the way. Similar projects are happening in other cities, including Singapore, Austin in Texas, Mountain View in California, and Ann Arbor in Michigan. In most of these cities, the cars will be fully autonomous but restricted to certain areas. The environments in which the vehicles work will gradually become broader and more complex, however.

One of the things driverless cars need is a set of highly detailed maps, and these are being developed for each of the pilot cities. The islands of precision mapping will then fan outwards from urban hubs along primary roads. Mapping firm TomTom says it has already covered 28,000 kilometres of roads in Germany with sufficient resolution for driverless cars – 4 per cent of all the roads in the country.

These inner-city vehicles are just one of two very different kinds of autonomous car that will be on our roads by 2018 or 2019. The Toyota Research Institute, led by Gill Pratt, calls these two types 'guardian angel' and 'chauffeur'.

The autonomous passenger-ferrying pods we will see in cities are chauffeurs. Guardian angels, on the other hand, are cars that will never fully wrest control from a human but will jump in to stop you doing something stupid. Both types of self-driving car should save lives. In the UK alone, 1,700 people are killed every year on the roads. Globally, the number is 1.25 million.

Sensors and software

Autonomous vehicles have many more eyes on the road than a human driver. Google's self-driving cars have eight sensors, Uber's driverless taxis 24, and Tesla's new cars will each have

21, all combining their data into a stream, rather as we integrate what our various senses are telling us.

These abilities are now being built into some cars as standard. For example, every car in Toyota's 2017 fleet – from the most basic model up – will have the sensors and software required to run in guardian angel mode. Sensors will enable automatic emergency braking, for example, allowing the car to stop itself if it detects an imminent collision.

Data from the sensors in all of its cars will be fed into Toyota's central data centre in Plano, Texas. Toyota's AI researchers will then use this to train their AIs how to drive on a wider variety of roads than are being considered in pilot schemes like the Gateway project. Ultimately, data gathered from guardian angel systems will help build cars that can drive as chauffeurs on any road. 'Our cars drive a trillion miles a year, which is a lot of data,' says Pratt.

Driverless cars could also have large social benefits, helping less mobile people get around. In places like Greenwich, the demographic set to increase the most over the next 20 years is people aged over 65. One problem that governments want to address is how older people can be cared for at home for longer. Using autonomous vehicles to drive them around could be a big help.

What will we do instead of driving?

Cars that are essentially chauffeurs are the vision of self-driving cars we are most familiar with from sci-fi films. They are also the vehicles that will change our relationship with cars the most. These vehicles don't need a steering wheel; they find their own way. Passengers can even ignore the outside world if they choose. In general, consumer expectations will dictate the way we use driverless cars. If people in cars want to work or relax or watch films, then car makers will cater to that.

If the pods in Greenwich are anything to go by, the size and shape of vehicles may change, too. It is not yet clear whether people would prefer to sit or stand, for example. Yet the width of roads and aerodynamics will still constrain what is possible. You can imagine completely different forms, but you probably cannot have a car 4 metres wide and 1 metre long, for example.

A car is an isolation chamber in which you sit for one or two hours a day. When we no longer need to drive one ourselves, we will be free to reimagine completely what we want to do with that time and space. In some ways, it might be helpful to compare a driverless car to a hotel or a house where the interior design is tailored to what people want from their experience – leisure, professional or tourism.

Safety will be the ultimate constraint on what we do inside our future vehicles. If autonomous software can be made so trustworthy that seat belts and airbags are no longer needed, the possibilities open up even further – sofas, beds ... you name it.

Watch where you're going, human!

When we share our roads with robots, how will we all get along? The trouble with humans is that we can be erratic. At the University of California, Berkeley, engineers are preparing autonomous cars to predict what we impulsive, unreliable humans might do next. A team led by Katherine Driggs-Campbell has developed an algorithm that can guess with up to 92 per cent accuracy whether a human driver will make a lane change.

Enthusiasts are excited that self-driving vehicles could lead to fewer crashes and less traffic. But people are not accustomed to driving alongside machines, says Driggs-Campbell. When we drive, we watch for little signs from other cars to indicate whether they might turn or change lanes or slow down. A robot might not have any of the same tics, and that could throw us off.

How do you ensure that driverless vehicles can communicate clearly with human drivers and pedestrians? Past algorithms have tried to predict what a human driver will do next by keeping tabs on body movements. If someone seems to be looking over their shoulder a lot, say, that might be a sign that they are thinking of changing lanes.

Driggs-Campbell and her colleagues wanted to see whether they could forecast a driver's actions by monitoring only outside the car. To see how human drivers do this, they asked volunteers to drive in a simulator. Each time the driver decided to change lanes, they pushed a button on the steering wheel before doing so. The researchers could then analyse data from the simulator for patterns at the time of lane changes: Where were all the cars on the road? How fast was each one going, and had it recently moved or slowed down? Was there sufficient room next to the driver's car?

They used some of the data to train the algorithm, then put the computer behind the wheel in reruns of the simulations. The algorithm could predict accurately when the driver would attempt a lane change. Such algorithms would help a self-driving car make smarter decisions in the moment. They could also be used to teach the cars to mimic human driving tics, says Driggs-Campbell.

Urban networking

For all the hype about self-driving cars, they are just one part of a larger revolution in urban transport. Increased use of artificial intelligence behind public transport services is easing our movement through the world in other ways. Both transitions are happening at the same time.

There is software that can knit together car pick-ups and the public transit network – sending an Uber car to meet you at just the right time to catch the train you need several kilometres away, for example. Pick your destination, press a button in an app, and simply follow the instructions to make your way across a city without having to think about it.

As large systems like Uber and Transport for London link up, getting a person from A to B starts to look like sending data across the Internet. That might prompt some of the same questions about fairness raised by the 'net neutrality' debate, which is concerned that all online traffic be treated equally. If you need to get to a meeting across London in 15 minutes, is it acceptable to pay a premium price that will change all the traffic lights to ensure you get no hold-ups on your route?

Cars with conscience

As our cars edge towards making these decisions for us, cases like these raise profound ethical questions. To drive safely in a human world, autonomous vehicles must learn to think like us – or at least understand how humans think. But how will they learn, and which humans should they try to emulate? These are tough questions. But people like ethicist Patrick Lin at California Polytechnic State University in San Luis Obispo insist that we cannot simply let manufacturers do what they want.

The ethical challenges raised by driverless cars can often be reduced to the trolley problem, a thought experiment familiar to philosophy students. Imagine a trolley car out of control, and five oblivious people on the track ahead. They will die if you do nothing – or you could flip a switch and divert the car to a different track where it will kill only one person. What should you do?

In a similar spirit, should an autonomous vehicle avoid a jay-walker who suddenly steps off the curb, even if it means swinging abruptly into the next lane? If a car that has stopped at an intersection for schoolchildren to cross senses a lorry approaching too fast from behind, should it move out of the way to protect the car's passengers, or take a hit and save the children? Such decisions may have to be programmed into the car – but what would we do in each case?

Answering such 'what do we do if …' questions is a two-step process. First, the vehicle needs to be able to accurately detect a hazard; second, it must decide on its response. The first step mainly depends on the efficient collection and processing of data on the whereabouts and speed of surrounding vehicles, pedestrians or other objects.

Moral drivers

Some hazards are clear, like not swerving into a river by the side of the road. Of course, not everything is that obvious. Consider the only death so far linked to driverless technology, which occurred in 2016. It happened because Tesla's autopilot system failed to detect that the whiteness ahead wasn't part of a bright spring sky but the side of a trailer. A human might have made that mistake, too, but sometimes, driverless vehicles make a mess of things we master intuitively. For example, one everyday situation autonomous cars have struggled with is someone walking behind a parked bus. A human mind would expect them to reappear, and supply a pretty accurate estimate of when and where − but, for a driverless car, that can be an extrapolation too far.

Even if a sensor system allows an autonomous car to assess its environment perfectly, the second step to driving in a morally informed way − taking the information gathered, assessing relative risks and acting accordingly − remains an obstacle course. At a basic level, it is about setting up rules and priorities. For example, avoid all contact with human beings, then animals, then property. But what happens if the car is faced with running over your foot or swerving into a building and causing millions of dollars' worth of damage?

Part of the problem with such a rules-based approach is that often there are no rules − at least, no single set that a sensor system based purely on obvious physical cues could hope to implement. For one thing, such a system cannot compute the societal cues we all rely on when driving. For another, the information a video camera or radar echo can supply is limited. Detecting a bus is one thing; detecting that it is full of school-children is more difficult.

Technologically, it's probably doable. A human intervention might program in details of the number and age of the passengers to be broadcast to surrounding vehicles, or sensors inside the bus might autonomously track its weight, including whether a person is sitting in a particular seat. But who decides a hierarchy of what lives are worth, and how do we eliminate discrimination and bias in how the cars are programmed?

Faking it

Getting computers to recognize other cars is surprisingly difficult. Although firms like Google and Uber are teaching their software by physically driving millions of miles in the real world, they also train their algorithms using pre-recorded footage of traffic. But there's a catch: computers need hundreds of thousands of laboriously labelled images, showing where vehicles begin and end, to make them expert vehicle recognizers. That takes people a lot of time and effort.

But it turns out that self-driving cars can be taught the rules of the road by studying virtual traffic in video games like *Grand Theft Auto V* (see Figure 4.2). Picking out cars in a video game is a similar task to doing it in reality, with the advantage that everything comes pre-labelled because it has been generated by the game's software. Matthew Johnson-Roberson at the University of Michigan in Ann Arbor and colleagues found that algorithms trained in the game were just as good at spotting cars in a pre-labelled data set as those trained on real roads. The video-game version needed around 100 times more training images to reach the same standard – but, given that 500,000 images can be generated from the game overnight, that is not a problem.

This is not the first time a research group has used video games to train artificial intelligence. Using simulations to train AI is starting to take off. In another example, Javier Romero at the Max Planck Institute for Intelligent Systems in Tübingen, Germany, and his colleagues are using fake people to help computers understand real human behaviour. The idea is that videos and images of computer-generated bodies walking, dancing and doing cartwheels could help computers learn what to look out for.

They have generated thousands of videos of 'synthetic humans' with realistic body shapes and movement. They walk, they run, they crouch, they dance. They can also move in less expected ways, but they are always recognizably human – and because the videos are computer-generated, every frame is automatically labelled with all the important information.

This lets AI learn to recognize patterns in how pixels change from one frame to the next, indicating how people are likely to move. Trained in this way, a driverless car can tell if a person is about to step into the road.

FIGURE 4.2 Games like *Grand Theft Auto V* can be realistic enough to help train AI.

Veil of ignorance

One way to avoid thorny moral questions is simply to ignore them. After all, a human driver is likely to know nothing about those in the vehicles around them. This 'veil of ignorance' approach amounts to developing responses to simple versions of likely situations – either by pre-programming them or letting the car learn on the job.

The first approach suffers from the problem that it is almost impossible to anticipate all possible scenarios. For example, in 2014 a Google car encountered a woman in an electric wheelchair chasing a duck into the road with a broom. The second approach seems more promising. A car might learn as it goes along, for example, that jaywalkers are more likely to be found on city streets than country roads, but that swerving to avoid one on a quiet country road brings less likelihood of hitting something else. Or it might learn that it's OK to break the speed limit occasionally to make way for an ambulance.

But basic rules still need to be programmed, and whole new ethical issues also arise: a programmer will not be able to predict what exactly a car will do in a given situation. We don't want autonomous vehicles to act unpredictably. Just as it's important for cars to predict the actions of human road users, so it matters for people to be able to anticipate a car's behaviour. Hence the question of what an autonomous car will do when it encounters that trolley-problem-like dilemma.

Some think that fixating on such an extreme case is not helpful. It is one possible situation in a million. It may be better to solve the more common problems – things like how to avoid pedestrians, stay within a lane, operate safely in bad weather, or push software updates to cars while safeguarding them from hackers. That may be so, but it misses what the thought experiment is about. Those who use it want to illustrate the point that car manufacturers alone do not have the moral authority to make all the decisions in their cars.

At the moment, companies such as Tesla and Google – which recently announced a withdrawal from building its own cars in favour of supplying software to other manufacturers – work on algorithms behind closed doors, but there are growing calls for transparency and common standards. In 2016 a US Department of Transportation team produced the first Federal Automated Vehicles Policy. It sets out decision-making ethics as one of 15 points the developers of autonomous vehicles should address, and calls on them to be transparent about their work on algorithms that 'resolve conflict situations'. It also urges companies to consult, to come up with solutions that are 'broadly acceptable'.

Similar calls are being made in the UK and Germany. Autonomous vehicles 'cannot be expected to make moral decisions around which society provides no agreed guidance', as Tim Armitage of the industry consortium UK Autodrive put it in a white paper compiled in 2017 by law firm Gowling WLG.

No solution will be perfect, warns Nick Bostrom, a philosopher at the University of Oxford and director of its Future of Humanity Institute. 'We should accept that some people will be killed by these cars.' Yet the challenge is not to build the perfect system but to build a system that is better than the one we have now – which kills more than a million people and injures around 50 million each year.

Highway code

Call them the Three Laws of Driverless Cars. In 2016 Germany's transport minister, Alexander Dobrindt, proposed a bill to provide the first legal framework for autonomous vehicles. It would govern how such cars perform in collisions where lives might be lost. The laws attempt to deal with what some call the 'death valley' of autonomous vehicles: the grey area between semi-autonomous and fully driverless cars that could delay the driverless future.

Dobrindt wants three things: for a car always to opt for property damage over personal injury; for a car never to distinguish between humans based on categories such as age or race; and for the car's manufacturer to be liable if there is a collision after a human removes his or her hands from the steering wheel – to check email, say.

'The change to the road traffic law will permit fully automatic driving,' says Dobrindt. He wants to put entirely autonomous cars on an equal legal footing to human drivers.

Lack of clarity about who is responsible for the operation of such vehicles is a major point of confusion among manufacturers, consumers and lawyers. In the United States, guidelines for companies testing driverless cars state that a human must keep their attention on the road at all times. This is also an assumption behind UK insurance for driverless cars, introduced in 2016, which stipulates that a human 'be alert and monitoring the road' at every moment. But that is clearly not what many people have in mind when thinking of driverless cars. 'When you say "driverless cars", people expect driverless cars,' says Natasha Merat at the University of Leeds, UK. 'You know – no driver.'

Dobrindt and others favour a ten-second rule, which requires a human to be sufficiently alert to take control of the vehicle within ten seconds. Similarly, Mercedes may require a driver to touch the wheel several times a minute. But ten seconds may not be enough. Just because you've put your hands back on the wheel does not mean you are in control of the vehicle. Merat has found that people can need up to 40 seconds to regain focus, depending on what they were doing at the time. Because of the lack of clarity, Merat thinks some car makers will wait until vehicles can be fully automated, without any human input whatsoever.

Driverless cars may end up being a form of public transport – like the driverless pods now being introduced in certain cities – rather than vehicles you own, says Ryan Calo at Stanford University, California. That would go down poorly in the United States, however. 'The idea that the government would take over driverless cars and treat them as a public good would get absolutely nowhere here,' says Calo.

AI doctors

Machines have already transformed health care. MRI scanners can peer inside the body and blood samples are analysed automatically, but human skill has always been an integral part of the process: a scan reveals a shadow and the oncologist recognizes its significance. But software could soon be working out what's wrong with you based only on medical data.

Doctors are often busy and overworked. They can make mistakes or overlook telltale symptoms. If computers could understand health on their own terms, perhaps they could speed up diagnosis and even make it more accurate. Take breast cancer detection. Diagnosis often requires information from three sources: an X-ray, an MRI scan and ultrasound. Cross-referencing is laborious and time-consuming – unless you use deep learning.

Researchers at Tel Aviv University in Israel have been using deep learning to analyse chest X-rays. Their system can distinguish between enlarged hearts and fluid build-up around the lungs. Meanwhile, a group at the National Institutes of Health Clinical Center in Bethesda, Maryland, is using similar methods to detect cancerous growths on the spine. IBM's Watson has turned its hand to diagnosis as well. In one case, it took just minutes to spot signs of a rare form of secondary leukaemia in a patient. This disease could otherwise have taken weeks to diagnose. And Google's DeepMind has several medical projects on the go as well, including the detection of early signs of eye disease.

DeepMind is collaborating with the UK's National Health Service to get access to large volumes of patient data. For example, through one partnership with Moorfields Eye Hospital in London, DeepMind is able to feed around a million anonymized retinal scans into its AI. This project will target two of the most common eye diseases – age-related macular degeneration and diabetic retinopathy. More than 100 million people around the world have these conditions.

The information that Moorfields is providing includes scans of the back of people's eyes, as well as more detailed scans known as optical coherence tomography (OCT). The idea is that the images will let DeepMind's neural networks learn to recognize subtle signs of degenerating eye conditions that even trained clinicians have trouble spotting. This could make it possible for a machine-learning system to detect the onset of disease before a human doctor could.

Gadi Wollstein, an eye doctor at the University of Pittsburgh, and his colleagues explored the use of neural networks to diagnose eye disease in 2005. But the team had a far smaller data set than DeepMind has been given. Wollstein says that a large data set is critical, as it allows the neural network to learn to recognize eye disease more completely and accurately.

Data overload

Ophthalmologists are using highly detailed OCT scans more and more. But this can lead to data overload. It's often hard for a doctor to see clear patterns and make good diagnoses, says Wollstein. He thinks a machine might do a better job. Any automated diagnosis software DeepMind comes up with could also make its way to high-street opticians, who are increasingly using OCT, says Pearse Keane, the Moorfields ophthalmologist who approached DeepMind in 2015.

DeepMind's partnership with Moorfields gives us an early look at how the marketplace for machine learning could work. DeepMind will not be paid for any of the work it does with the Royal Free or Moorfields Eye Hospitals. However, it is able to test out algorithms on real data sets that describe serious problems, and it gets to keep the neural networks it trains using that data. The valuable knowledge about eye disease contained in Moorfields' anonymous data set will become the property of DeepMind, built into its artificial intelligence systems. In effect, training its machine-learning systems on real-world health data is DeepMind's payment for advancing the field of diagnostic AI.

But will doctors – or patients – ever accept the word of a machine? Deep learning's complex networks are inscrutable, typically spitting out conclusions without giving reasons. For example, if you have ever had Facebook suggest you tag someone you don't know as one of your friends, not even a Facebook engineer could tell you why that happened. Apply that level of mystery to medicine and people may well become uneasy.

One way to get clinicians to embrace such systems is to use the outputs of the deep-learning software to train another, transparent model whose answers humans can inspect and understand. Work in this area is going to be as much about people – and what we will come to accept – as it is about AI.

Killer robots

One of the most heated debates in AI concerns the development of so-called killer robots. Advocates of autonomous weapons argue that wars fought by machines instead of humans would be more humane. Human rights abuses committed by soldiers during conflict are certainly all too common, but could machines really do better? While many think this is an outrageous idea, others think not only that machines could but that they must. 'Humans are currently slaughtering other humans unjustly on the battlefield,' says Ronald Arkin, a robotics expert at the Georgia Institute of Technology in Atlanta. 'I can't sit idly by and do nothing. I believe technology can help.'

The development of lethal autonomous weapons systems – killer robots – is accelerating, with many of the world's armies looking for ways to keep their soldiers out of the line of fire. Sending robots in place of human soldiers would save lives, particularly for the nation possessing such advanced technology. And unlike humans, robots will not break the rules.

The issue is also rising up the international agenda. In the past few years, the United Nations has discussed so-called lethal autonomous weapons systems many times. But, with strident opposition coming from groups such as the Campaign to Stop Killer Robots, there are signs that the discussions are becoming more urgent. Nine nations have called for a ban on lethal autonomous weapons systems and many others have stated that humans must retain ultimate control of robots.

Robots already play several roles on the battlefield. Some carry equipment, others dismantle bombs and still others provide surveillance. Remote-controlled drones let their operators control attacks on targets from thousands of kilometres away. The latest machines, though, take drones to the next level. Capable of selecting and engaging targets with little or no human intervention, the authorization to open fire is sometimes all that remains under human control.

The US Navy's Phalanx anti-missile system on board its Aegis ships can perform its own 'kill assessment' – weighing up the likelihood that a target can be successfully attacked. UK firm BAE is developing a crewless jet called Taranis. It can take off, fly to a given destination and identify objects of interest with little intervention from ground-based human operators unless required. The jet is a prototype and carries no weapons, but it demonstrates the technical feasibility of such aircraft. Meanwhile, Russia's 'mobile robotic complex' – a crewless tank-like vehicle that guards ballistic missile installations – and South Korea's Super Aegis II gun turret are reportedly able to detect and fire on moving targets without human supervision. The Super Aegis II can pinpoint an individual from 2.2 kilometres away.

Arms manufacturers dislike talking about the details. The specifics, in general, are classified. What is clear, though, is that technology is no longer the limiting factor. In the words of a spokesman for UK missile manufacturers MBDA, 'technology is not the likely restriction as to what is feasible in the future'. Instead, autonomous weapons will be constrained by policy, not capability.

Starfish killer

In 2016 robots started to shoot to kill, no questions asked. This wasn't a *RoboCop* (see Figure 4.3) remake but real life on Australia's Great Barrier Reef, where a killer robot is being deployed against coral-wrecking starfish. Called COTSbot, it is one of the world's most advanced autonomous weapons systems, capable of selecting targets and using lethal force without any human involvement.

A starfish-killing robot may not sound like an internationally significant development, but releasing it on to the reef would cross a Rubicon. COTSbot amply demonstrates that we now have the technology to build robots that can select their own targets and autonomously decide whether to kill them. The potential applications in human affairs – from warfare to law enforcement – are obvious.

Against this background, COTSbot is a good thing – a chance to test claims about autonomy, accuracy, safety, hackability and so on in a relatively benign environment. It also offers an opportunity to demonstrate that autonomous robots can do good as well as bad. But the real significance is that it shows that *RoboCop* is getting ever closer to reality.

Rules of engagement

What, then, are the relevant rules of war? There are no laws specifically covering robots, but all weapons must comply with existing conventions. One key principle is that civilians and civilian property must not be intentionally targeted. Weapons must also be capable of discriminating between civilians and soldiers. And the use of force must be proportional – the expected military advantage of an attack cannot be outweighed by collateral damage.

Within such a framework, Ronald Arkin believes that, if a lethal autonomous system can be shown to be better than a human soldier at limiting civilian casualties, then a ban on such technology would be a mistake. 'We have to remember the fallibility and frailty of humans in modern warfare,' he says. 'If we can do better than them, we are saving lives.'

Others make the case just as strongly. Erik Schechter at Vanderbilt University in Nashville, Tennessee, has argued in *The Wall Street Journal*, for example: 'If the goal of international humanitarian law is to reduce non-combatant suffering in wartime, then using sharpshooting robots would be more than appropriate, it would be a moral imperative.'

Robots could spare soldiers' lives, too. Instead of calling in an airstrike to bomb a suspected enemy base, which might be situated in a densely populated urban area, for example, robots could enter the building ahead of human soldiers, taking the

FIGURE 4.3 Killer robots, like this one in the 1988 film *RoboCop*, are familiar from science fiction – how will the real ones differ?

initial risks. Where aspects of any mission are especially risky, machines could take the lead.

The right to dignity

The idea is deeply divisive, however. For many, the prospect of a computer chip having the power of life or death over someone is discomforting. According to Christof Heyns, the UN's Special Rapporteur on extrajudicial, summary or arbitrary executions, it could contravene humanitarian law and the human right to dignity.

'Humans need to be quite closely involved in the decision for it not to violate your human dignity,' says Heyns. He notes that a robot's targets don't have the option of an appeal to humanity the way they might if a person was behind the weapon. It would be like being exterminated, he says. Remote-controlled drones already give little opportunity to make such appeals. But they at least have a human operator, however distant, who can make ethical judgements. 'The hope that this is possible is at least not completely absent,' says Heyns. 'And hope is part of a dignified life.'

Ultimately, Heyns is wary of what he calls the 'depersonalization of force'. In a 2013 report to the UN, he warned that 'tireless war machines, ready for deployment at the push of a button' could lead to a future of permanent conflict. If governments don't have to put boots on the ground, going to war could become too easy. Even in scenarios where machines fight machines, significant collateral damage could destroy a nation's infrastructure. And as casualty numbers will be lower, wars could go on for longer, preventing post-war reconstruction.

Noel Sharkey, an AI and robotics researcher at the University of Sheffield, UK, and a leading member of the Campaign to Stop Killer Robots, has been trying to bring the issue to international attention for nearly a decade. A key driver of his dogged campaigning is his awareness of the shortcomings of current technology. While Arkin has an eye on next-generation technology, Sharkey is concerned with the present. 'I could make you a killer robot within weeks that detects human body signatures and fires at them,' he says. 'The problem is being able to discriminate between a civilian and a combatant.'

Bottling that ability is hard. Aralia Systems is a UK-based firm that provides image-analysing software for security applications. It can highlight suspicious activity in CCTV footage, for example. In 2015 the system flagged up the activity of a group of people who were later found to have been scoping out a public area for a suitable place to plant a bomb. The individuals were apprehended and successfully prosecuted, says Wright. But the company's co-founder, Glynn Wright, readily admits that making snap decisions in busy urban environments is a long way off.

Interview: Should we ban autonomous weapons?

Mark Bishop is Professor of Cognitive Computing at Goldsmiths, University of London, and chairs the Society for the Study of Artificial Intelligence and the Simulation of Behaviour. In this 2013 New Scientist *interview he explains why banning weapons that can deploy and destroy without human intervention is vital.*

What is the Campaign to Stop Killer Robots?

It is a confederation of NGOs and pressure groups lobbying for a ban on producing and deploying fully autonomous weapon systems – where the ability of a human to both choose the precise target and intervene in the final decision to attack is removed.

How close are we to this?

Examples already exist. Some, such as the Phalanx gun system, used on the majority of US Navy ships to detect and automatically engage incoming threats, have been around for some time. Another is the Israeli Harpy 'fire-and-forget' unmanned aerial vehicle, which will seek out and destroy radar installations.

What's driving the technology's development?

Current Western military strategy focuses more on drones than on traditional forces, but remote-controlled drones are vulnerable to hijacking. Fully autonomous systems are virtually immune to this. They also lower costs. This means manufacturers sell more, so there is a commercial imperative to develop autonomous systems and for governments to deploy them.

What are the dangers?

There are reasons to doubt whether autonomous systems can appropriately judge the need to engage, react to threats proportionately or reliably discriminate between combatants and civilians. Also, when you get complex software systems interacting, there is huge potential for unforeseen consequences. A vivid example was seen on Amazon in 2011 when pricing bots raised the cost of a book, *The Making of a Fly*, to over $23 million.

Are you worried about escalation?

Yes. In South Korea scientists are developing a robot to patrol the border with North Korea. If this were deployed and incorrectly or disproportionately engaged, it is easy to imagine a minor border incursion escalating into a serious confrontation. Even more frighteningly, in 1983, during the US military exercise Able Archer, Russian automatic defence systems falsely detected an incoming missile and it was only a Russian colonel's intervention that averted nuclear war. But the potential for escalation gets particularly scary when you have autonomous systems interacting with other autonomous systems.

Couldn't robots reduce risk to humans?

There is a case, put forward by people such as US roboticist Ronald Arkin, that robots might make more dispassionate assessments than grieving or revenge-seeking soldiers. Not only does this not address the problem of escalation but it also only holds water if systems can reliably decide when to engage, judge proportionality and accurately discriminate targets.

So what should we do?

The technology behind autonomous systems has other uses, such as the Google car-driving system, so banning development would be difficult. Instead, we must focus on a global treaty banning the deployment of autonomous weapons.

Calling the shots

Even if a machine were capable of discriminating between targets and civilians, what of its ability to make moral decisions based on such data? Arkin has argued that we could develop software that acts as an 'ethical governor' to guide a robot's response to various circumstances. The required complexity of such software, though, means that these proposals remain on the drawing board. For now, a human must remain in control.

But what does that mean? The debate on killer robots turns on this question and the call for a ban will succeed or fail depending on what counts as having a human making decisions. There is no accepted definition. And there is another issue: focusing only on the question of killer robots conceals more fundamental problems with the way humans and machines interact. Just because a human is involved does not mean that the problems with high-tech killing go away.

In 2003, for example, an operator at a US Patriot missile battery received an automated alert that an incoming Iraqi missile had been detected. She had a split second to make a decision and chose to take defensive action, authorizing the battery to fire. But the target turned out to be an RAF Tornado jet, whose two pilots were killed when the Patriot missile hit. What went wrong? The system certainly misidentified the jet. But an inquest found that this was because the operators hadn't been properly trained. As a result, the system had not been connected to a wider network that would have told it that the jet was in contact with air traffic controllers and not a threat. In this case, having humans in the loop was arguably the root of the problem.

Whatever the outcome in the UN, the issues raised in the debate over autonomous weapons systems are less simple than they first seem. Whether fully autonomous or not, machines are already part of war and the promise to keep people involved is certainly no guarantee that mistakes won't happen. Machines don't streamline war; they complicate it. It's something they have in common with us.

5
Into the unknown

*How computers may overcome the limitations of
the human mind*

*Right now, computers are mostly augmenting human intelligence. But
some machines are beginning to solve problems without being limited by
the human mind, allowing them to invent novel gadgets or even push back
the frontiers of mathematics. Will machines one day supplant thinkers and
tinkerers altogether? Even where they outperform us, their discoveries will
only be useful if we can make sense of them and apply them.*

Eureka machines

We are used to serendipity and invention going hand in hand. Take the dawn of mechanical flight, for example. On a summer's day in 1899 a bicycle mechanic in Dayton, Ohio, slid a new inner tube out of its box and handed it to a customer. The pair chatted and the mechanic toyed idly with the empty box, twisting it back and forth. As he did so, he noticed the way the top of the box distorted in a smooth, spiral curve. It was a trivial observation – but one that would change the world.

The shape of the box just happened to remind the mechanic of a pigeon's wing in flight. Watching that box flex in his hands, Wilbur Wright saw how simply twisting the frame supporting a biplane's wings would give him a way to control an aircraft in the air.

The Wright brothers' plane is just one of many examples. Another is Velcro: George de Mestral invented the material after he noticed the hook-covered seeds of the burdock plant sticking to his dog. And Harry Coover's liquid plastic concoction failed miserably as a material for cockpit canopies, as it stuck to everything. But it had a better use: superglue.

It may be romantic, but it is an achingly slow way to advance technology. Relying on happenstance means inventions that could be made today might not appear for years. 'The way inventions are created is hugely archaic and inefficient,' says Julian Nolan, CEO of Iprova, a company based in Lausanne, Switzerland, which specializes in generating inventions. Nothing has changed for hundreds of years, he says. 'That's totally out of sync with most other industries.'

But we are starting to make our own luck. Those eureka moments could soon be dialled up on demand as leaps of imagination are replaced by the steady steps of software. From algorithms that mimic nature's way of producing the best designs to systems that look for gaps between existing patented technologies that new designs might fill, computer-assisted invention is here.

The impact could be huge. Some claim that automated invention will speed up technological progress. It could also level the playing field, making inventors of us all. But what happens if the currency of ideas is devalued? To qualify for a patent, for example, an idea can't be 'obvious'. How does that apply when ideas are found by brute force?

What are genetic algorithms?

Genetic algorithms – also known as evolutionary algorithms – tackle the problem of design by mimicking natural selection (see Figure 5.1). Desired characteristics are described as if they were a genome, where genes represent parameters such as voltages, focal lengths or material densities, say.

The process starts with a more or less random sample of such genomes, each a possible, albeit suboptimal, design. By combining parent genomes from this initial gene pool and introducing 'mutations', offspring are created with features of each parent, plus potentially beneficial new traits. The fitness of the offspring for a given task is tested in a simulation. The best are selected and become the gene pool for the next round of breeding. This process is repeated again and again until, as with natural selection, the fittest design survives.

As well as evolving new designs, genetic algorithms can be used to evolve 'parasites' that inflict maximal damage to test safety or security features. 'Nature has been very good and very creative at finding loopholes in every possible complex system,' says Eric Bonabeau of Icosystem of Cambridge, Massachusetts, who has used this technique to improve the design of ships for the US Navy.

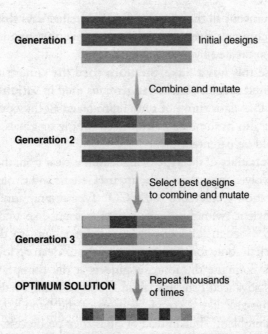

Generation 1 — Initial designs

Combine and mutate

Generation 2

Select best designs
to combine and mutate

Generation 3

OPTIMUM SOLUTION — Repeat thousands
of times

FIGURE 5.1 Genetic algorithms try to find the optimum solution
to a problem by repeatedly combining and mutating the best in each
generation of potential solutions.

As nature intended

The first group of researchers to mimic evolution in patent
design – pioneering the use of so-called genetic algorithms –
was led by John Koza at Stanford University in California in
the 1990s. The team tested their algorithms by seeing whether
they could reinvent some of the staples of electronic design: the
early filters, amplifiers and feedback control systems developed
at Bell Labs in the 1920s and 1930s. They succeeded. 'We were

able to reinvent all the classic Bell Labs circuits,' says Koza. 'Had these techniques been around at the time, the circuits could have been created by genetic algorithms.'

In case this was a fluke, the team tried the same trick with six patented eyepiece lens arrangements used in various optical devices. The algorithm not only reproduced all the optical systems but also in some cases improved on the originals, in ways that could be patented.

The versatility of this type of algorithm is clear from the showcase of evolved inventions at the annual Genetic and Evolutionary Computation Conference (GECCO). Typical innovations range from efficient swimming gaits for an octopus-like underwater drone to a design for low-power computer chips to the most fuel-efficient route for a future space probe to clean up low-Earth orbits. To compute the route, engineers at the European Space Agency's advanced concepts lab in Noordwijk, the Netherlands, treated the task like a cosmic version of the famous travelling salesman problem – but, instead of cities, their probe visits derelict satellites and dead rocket bodies to nudge them out of orbit.

However, the big prize at GECCO is the human competitiveness award, or 'Humie', for inventions deemed to compete with human ingenuity. The first Humie, in 2004, was awarded for an odd-shaped antenna, evolved for a NASA-funded project. It worked brilliantly even though it looked like a weedy sapling, with a handful of awkwardly angled branches rather than a regular stick-like antenna. It was certainly not something a human designer would produce.

That is often the point. 'When computers are used to automate the process of inventing, they aren't blinded by the preconceived notions of human inventors,' says Robert Plotkin, a patent lawyer in Burlington, Massachusetts. 'So they can produce designs that a human would never dream of.'

Into the unknown

There is just one problem with using genetic algorithms: you need to know in advance what you want to invent so that your algorithm can modify it in fruitful ways. Genetic algorithms tend to be good at optimizing pre-existing inventions but not so good at coming up with things that are genuinely novel. This is because they don't take big, inventive leaps. It also means that they have less chance of making a commercially valuable hit.

One approach is to use software to help inventors notice easily missed features of a problem that, if addressed, could lead to a novel invention. 'An invention is something new that was not invented before because people overlooked at least one thing that the inventor noticed,' says Tony McCaffrey, chief technology officer of Innovation Accelerator based in Natick, Massachusetts. 'If we can get people to notice more obscure features of a problem, we raise the chances that they will notice the key features needed to solve it.'

To do that, Innovation Accelerator has written software that lets you describe a problem in human language. It then 'explodes' the problem into a large number of related phrases and uses these to search the US Patent and Trademark Office database for inventions that solve similar problems. The system is designed to look for analogues to the problem in other domains. In other words, the software does your lateral thinking for you.

In one example, McCaffrey asked the system to come up with a way to reduce concussion among American football players (see Figure 5.2). The software exploded the description of the problem and searched for ways to reduce energy, absorb energy, exchange forces, lessen momentum, oppose force, alter direction and repel energy. Results for how to repel energy led the firm to invent a helmet that contained strong magnets to repel other players' helmets, lessening the impact of head

FIGURE 5.2 A software system came up with a novel helmet design to protect American football players from concussion.

clashes. Unfortunately, someone else beat them to the patent office by a few weeks. But it proved the principle.

In another case, the software duplicated a ski maker's recent innovation. The problem was to find a way to stop skis vibrating so skiers could go faster and turn more safely. The manufacturer eventually stumbled upon an answer, but Innovation Accelerator's software was able to find it quickly. 'A violin builder had a method to produce purer music by reducing vibrations in the instrument,' says McCaffrey. 'The method was applied to the skis and made them vibrate less.'

Trend spotting

The technology at Nolan's firm, Iprova, also helps inventors to think laterally – but with ideas derived from sources far beyond patent documents. The company is unwilling to reveal exactly how its Computer Accelerated Invention technique works, but in a 2013 patent Iprova says it provides clients with 'suggested innovation opportunities' by interrogating not only patent databases and technical journals but also blogs, online news sites and social networks.

Of particular interest is the fact that it alters its suggestions as tech trends on the Internet change. The result seems to be extremely productive. The company uses its technology to create hundreds of inventions per month, which its customers can then choose to patent. If their wide range of customers in the health-care, automotive and telecommunications industries is anything to go by, Iprova appears to have been successful. One of its clients is Philips, a major technology multinational. Such firms don't add outside expertise to their research and development teams lightly.

All this means that algorithm-led discovery is likely to be the most productive inventing process of the future. 'Human inventors who learn to leverage computer-automated innovation will leapfrog peers who continue to invent the old-fashioned way,' says Plotkin. But where do we draw the line between the two? It may turn out that there is no clear separation between human and algorithm, with the key being to find the right division of labour. However, if the division of labour is too much on the computer's side, it could undermine the patent system itself. Currently, a 'person having ordinary skill in the art' must believe that an invention is not obvious if it is to be granted a patent. But if inventors are only tending a computer, the inventions that arise could be deemed an obvious output of that computer, like hot water from a kettle.

Chance favours the prepared mind, the saying goes. If Wilbur Wright had not been thinking about his aircraft while serving a customer, he may never have had his eureka moment. Creative software could make such accidental juxtapositions far more common. 'Outsource serendipity to the algorithm,' says Bonabeau.

What's for dinner?

Find yourself cooking the same tried-and-trusted favourites, using the same old ingredients? The Chef Watson app could help. It uses the brain of IBM's Watson supercomputer to invent new dishes.

The key to the Chef Watson app is the supercomputer's ability to devour large amounts of information and make links between chunks of it. It has already proved its mettle by winning the quiz show *Jeopardy!* and is being used to help doctors make cancer diagnoses at Memorial Sloan Kettering Hospital in New York. Now it is attempting to do something arguably more difficult: using creativity to invent recipes people will actually want to try.

To provide the data, IBM teamed up with US recipe website Bon Appetit. This has a database of more than 9,000 recipes, tagged according to their ingredients, type of dish and cooking style – Cajun or Thai, for example. Watson creates a statistical correlation between ingredients, styles and recipe steps in the database and uses this to work out which ingredients usually go together and what each type of food requires. 'That's how it knows that a burrito, a burger and a soup all need different elements,' says IBM's Steve Abrams. 'It knows a burrito always needs a wrapper of some sort, whereas soup always needs liquid. That's how you don't get a runny burrito.'

To use the app, the cook first types in an ingredient they want to use. Next, they decide the amount of experimentation

---

they want Watson to take, ranging from 'Keep it classic' to 'Surprise me'. Watson then offers the cook further ingredients, styles and dishes that it thinks usually go well with the initial ingredient. They can promote or exclude ingredients by clicking the 'Love it' or 'Hate it' buttons. Finally, the cook hits search, and Watson analyses its database to come up with a range of basic recipes that can then be further tweaked to make them more or less experimental.

If the cook wants to ramp it up, a database of flavour compounds found in a wide range of foods is also consulted and used to combine ingredients that should theoretically fit together, such as vodka and Brussels sprouts, or cauliflower and chrysanthemum. Psychological research into what flavours people find more or less pleasant is taken into account, as is a 'surprise' score for combinations of ingredients: the higher the score the less often they are found in recipes together.

As you move to the experimental end of the scale, you are looking less at what ingredients go nicely together and more at the flavour compounds that the ingredients have in common. 'It works out what things might go together that you'd never think of,' says Abrams. That's the theory, at least. In practice, Chef Watson can suggest some odd substitutions. In a creamy pasta dish, crème fraiche might be replaced by a glass of milk. Turn the setting up to 'Surprise me' and Watson might try to persuade you that your tuna bake needed half a kilo of goose meat. It can occasionally be inspired, however. A suggestion to replace clam juice in a classic gumbo dish with the Japanese soup base dashi works, giving the dish a deep savoury flavour that you may never have thought to try out.

The plan is for Chef Watson to become more sophisticated and to suck in data from more sources. It already scours Wikipedia's world cuisine pages and looks up nutrition facts in a US Department of Agriculture database for help on determining ingredient proportions.

Apart from the occasional odd ingredient combination, there are a few other bugs in the way Chef Watson conjures creative recipes out of pure data. One tester found that it seemed to struggle with portion sizes. Another found that a recipe called for precisely 554 juniper berries, and she was also told to 'skin and bone' her tofu.

Proof of concept: software to crack mathematical problems

Pure mathematics is another human endeavour characterized by leaps of inspiration. Software is already cracking theorems that humans have struggled to prove. Could computers take on a more creative role, too? If so, future computers could take mathematics to places too complex for our brains to follow.

In 2012 Shinichi Mochizuki, a highly respected mathematician at Kyoto University in Japan, published more than 500 pages of dense maths on his website. It was the culmination of years of work. Mochizuki's inter-universal Teichmüller theory described previously uncharted areas of the mathematical realm and let him prove a long-standing conundrum about the true nature of numbers, known as the ABC conjecture. Other mathematicians hailed the result, but warned that it would take a lot of effort to check. Months passed with no conclusion. In the end, it took four years for anyone to start making sense of it.

Ask a mathematician what a proof is and they're likely to tell you it must be absolute – an exhaustive sequence of logical steps leading from an established starting point to an undeniable conclusion. But that's not the whole story. You cannot just publish something you believe is true and move on – you have to convince others that you have not made any mistakes. For a truly groundbreaking proof, this can be a frustrating experience.

It turns out that very few mathematicians are willing to put aside their own work and dedicate the months or even years it would take to understand a proof like Mochizuki's. And as maths becomes increasingly fractured into subfields within subfields, the problem is set to get worse. Some think maths is reaching a limit. Real breakthroughs can be too complicated for others to check, so many mathematicians occupy themselves with more attainable but arguably less significant problems. What is to be done?

For some, the solution lies in employing digital help. Many mathematicians already work alongside computers – they can help check proofs and free up time for more creative work. But it might mean changing how maths is done. What is more, computers may one day make genuine breakthroughs on their own. Will we be able to keep up? And what does it mean for maths if we can't?

Four colours

The first major computer-assisted proof was published 40 years ago and it immediately sparked a row. It was a solution to the four-colour theorem, a puzzle dating back to the mid-nineteenth century. The theorem states that all maps need only four colours to make sure no adjacent regions are coloured the same. You can try it as many times as you like and find it to be true (see Figure 5.3). But to prove it, you need to rule out the very possibility of there being a bizarre map that bucks the trend.

In 1976 Kenneth Appel and Wolfgang Haken did just that. They showed that you could narrow the problem down to 1,936 sub-arrangements that might require five colours. They then used

FIGURE 5.3 It should be possible to colour in any map using just four colours and with no two adjacent regions in the same colour. Try it for yourself with the 'map' above, devised in the 1970s by the writer and maths popularizer Martin Gardner.

a computer to check each of these potential counter-examples, and found that all could indeed be coloured with just four colours.

Job done, or so you might think, but mathematicians were reluctant to accept this as a proof. What if there was an error in the code? They did not trust the software – but nobody wanted to double-check the thousand particular cases by hand. They had a point. Checking software that tests a mathematical conjecture can be harder than proving it the traditional way and a coding mistake can make the results totally unreliable.

The trick is to use software to check software. Working with a type of program known as a proof assistant, mathematicians can verify that each step of a proof is valid. The process is interactive. You type commands into the tool and then the tool checks it – much like a spellchecker. And what if the proof assistant has a bug? That is always possible, but these programs tend to be small and relatively easy to check by hand. What's more, as the code is run over and over again, you gain evidence to show that it is computing properly.

Boring details

However, using proof assistants means embracing a different way of working. When mathematicians write out proofs, they skip a lot of the boring details. There is no point in laying out the foundations of calculus every time, for example. But such shortcuts don't work with computers. To work with a proof, they must account for every logical step, even apparent no-brainers such as why $2 + 2 = 4$.

Translating human-written proofs into computer-speak is still an active area of research. A single proof can take years. One early breakthrough came in 2005, when Georges Gonthier at Microsoft Research Cambridge, UK, and his colleagues updated the proof of the four-colour theorem, making every part of it computer-readable. Previous versions, ever since Appel and Haken's work in 1976, relied on an area of maths called graph theory, which draws on our spatial intuition. Thinking about regions on a map comes naturally to humans, but not computers. The whole thing needed reworking.

Gonthier discovered that a part of the proof – widely assumed to be true because it seemed so obvious – had in fact never been proved at all because it was deemed not worth the effort. The assumption turned out to be correct, but it illustrates an added benefit of extra precision. 'You have to turn everything into algebra, and that forces you to be more precise,' he says. 'That precision ends up paying off.'

Tackling the four-colour theorem was just the beginning, however. 'It has relatively few uses in the rest of mathematics,' says Gonthier. 'It was a brain-teaser.' So he turned to the Feit–Thompson theorem, a large and foundational proof in group theory from the 1960s. For many years the proof had been built upon and rewritten and it was eventually published in two books. By formalizing it, Gonthier hoped to demonstrate the computer's capacity to digest a meatier proof that touched many different branches of mathematics. It was the perfect test case.

Unaccepted proofs

It was a success: in the process, they found a couple of minor mistakes in the books. These were easily fixed but they were still things that every human mathematician had missed. People took notice, says Gonthier. 'I got letters saying how wonderful it was.' In both cases, the result was never in doubt. Gonthier was taking well-established maths and translating it for computers. But others have been forced to redo their work in this way just to get their proofs accepted.

In 1998 Thomas Hales at the University of Pittsburgh, Pennsylvania, found himself in a similar position to the one Mochizuki is in today. He had just published a 300-page proof of the Kepler conjecture, a 400-year-old problem that concerns the most efficient ways to stack a collection of spheres. As with the four-colour theorem, the possibilities boiled down to variations on a few thousand arrangements. Hales and his student Samuel Ferguson used a computer to check them all.

Hales submitted his result to the journal *Annals of Mathematics*. Five years later, reviewers for the journal announced they were 99 per cent certain that the proof was correct. 'Referees in mathematics generally do not want to check computer code. They don't see that as part of their job,' says Hales. Convinced he was right, Hales began to rework his proof in 2003, so that it could be checked with a proof assistant. It essentially meant starting all over again. It took him more than ten years to complete the project.

Gonthier's and Hales' research has shown that computers can help make progress in important areas of mathematics. 'The big theorems in maths that we're proving now seemed a distant dream ten years ago,' says Hales. But, despite advances like the proof assistant, proving things with a computer is still a laborious process. Most mathematicians don't bother.

That is why some are working in the opposite direction. Rather than making proof assistants easier to use, Vladimir Voevodsky at the Institute for Advanced Study in Princeton, New Jersey, wants to make mathematics more amenable to computers. To do this, he is redefining its very foundations.

True to type

This is deep stuff. Maths is currently defined in terms of set theory, essentially the study of collections of objects. For example, the number zero is defined as the empty set, the collection of no objects. One is defined as the set containing one empty set. From there, you can build an infinity of numbers. Most mathematicians don't worry about this on a day-to-day basis and take it for granted that they will understand each other without going down to that much detail.

This is not the case for computers, and that's a problem. There are multiple ways to define certain mathematical objects in terms of sets. For us, that doesn't matter, but if two computer proofs use different definitions for the same thing, they will be incompatible. 'We cannot compare the results, because at the core they are based on two different things,' says Voevodsky. 'The existing foundations of maths don't work very well if you want to get everything in a very precise form.'

Voevodsky's alternative approach swaps sets for types – a stricter way of defining mathematical objects in which every concept has exactly one definition. Proofs built with types can also form types themselves, which is not the case with sets. This lets mathematicians formulate their ideas with a proof assistant directly, rather than having to translate them later. In 2013 Voevodsky and colleagues published a book explaining the principles behind the new foundations. In a reversal of the norm, they wrote the book with a proof assistant and then 'unformalized' it to produce something more human-friendly.

This backwards way of working changes the way mathematicians think, says Gonthier. It also allows much closer collaboration between large groups of mathematicians, because they don't have to constantly check each other's work. In turn, this has started to popularize the idea that proof assistants can be good for the working mathematician.

It may be just the beginning. By making maths easer for computers to understand, Voevodsky's redefinition might take us into new territory. As he sees it, mathematics is split into four quadrants (see Figure 5.4). Applied maths – modelling the airflow over a wing, for example – involves high complexity but low abstraction. Pure maths, the kind of pen-and-paper maths that is far removed from our everyday lives, involves low complexity but high abstraction. And school-level maths is neither complex nor abstract. But what lies in that fourth quadrant?

Left behind

'It is very difficult at present to go into the high levels of complexity and abstraction, because it just doesn't fit into our heads very well,' says Voevodsky. 'It somehow requires abilities that we don't possess.' By working with computers, perhaps humans could access this fourth mathematical realm. We could prove bigger, bolder and more abstract problems than ever before, pushing our mastery of maths to ultimate heights.

Alternatively, perhaps we will be left behind. In 2014 Alexei Lisitsa and Boris Konev at the University of Liverpool, UK, published a computer-assisted proof so long that it totalled 13 gigabytes, roughly the size of Wikipedia. Each line of the proof is readable, but for anyone to go through the entire result would take several tedious lifetimes.

The pair have since optimized their code and reduced the proof to 800 megabytes – a big improvement, but still impossible to digest. From a human viewpoint, there is little difference. Even if you did devote your life to reading something like this, it would be like studying a photograph pixel by pixel, never seeing the larger picture. 'You cannot grasp the idea behind it,' says Lisitsa.

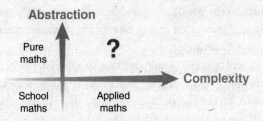

FIGURE 5.4 Maths that is both highly abstract and highly complex may be beyond human ability. Some think computers could open up this new territory for us.

Although it is on a far grander scale, the situation is similar to the original proof of the four-colour theorem, where mathematicians could not be sure an exhaustive computer search was correct. 'We still don't know why the result holds true,' says Lisitsa. 'It could be a limit of human understanding, because the objects are so huge.'

Doron Zeilberger of Rutgers University in Newark, New Jersey, thinks that there will even come a time when human mathematicians will no longer be able to contribute. 'For the next hundred years humans will still be needed as coaches to guide computers,' he says. 'After that, they could still do it as an intellectual sport, and play each other like human chess players still do today, even though they are much inferior to machines.'

Zeilberger is an extreme case. He has listed his computer, nicknamed Shalosh B. Ekhad, as a co-author for decades and thinks humans should put pen and paper aside to focus on educating our machines. 'The most optimal use of a mathematician's time is knowledge transfer,' he says. 'Teach computers all their tricks and let computers take it from there.'

Spiritual discipline

Most mathematicians, however, bristle at the idea of software that churns out proofs beyond human comprehension. 'The idea that computers are going to replace mathematicians is misplaced,' says Gonthier.

Besides, computer mathematicians would risk churning out an accelerating stream of unread papers. As it stands, scientific results often fail to garner the recognition they deserve, but the problem is particularly marked for maths. In 2014 more than 2,000 maths papers were posted to the online repository arXiv.org each month, more than in any other discipline, and the rate is increasing. With so many new results appearing, many go unnoticed. One option would be to create software that reads everything published and helps humans keep up with the important stuff.

But Gonthier feels this is missing the point. 'Mathematics is not as much about finding proofs as it is about finding concepts,' he says. The nature of maths itself is under scrutiny. If humans do not understand a proof, then it doesn't count as maths, says Voevodsky. 'The future of mathematics is more a spiritual discipline than an applied art. One of the important functions of mathematics is the development of the human mind.'

All of this may be too late for Shinichi Mochizuki, however. His work is so advanced, so far removed from mainstream maths, that having a computer check it would be far more difficult than coming up with the original proof. 'I don't even know if it would be possible to formalize what he's done,' says Hales. For now, humans remain the ultimate judge – even if we don't always trust ourselves.

6
Machines that create

The AI world of art and storytelling

Machines are taking their first steps into the world of art — learning how to tell stories, compose music and paint pictures by themselves. We would not hesitate to call these activities creative when done by a human. Will we ever be comfortable saying the same of a machine?

Plot bots: AI storytelling

What if there were a monkey that was afraid of bananas? What if a man woke up as a dog but could still use his phone? What if there were a house without a door? The What-if Machine has an active imagination, much like us: we love to make things up. We tell stories to entertain, to share experiences and to make sense of things. As the author Philip Pullman put it, 'After nourishment, shelter and companionship, stories are the thing we need most in the world.'

Soon, though, we won't be the only ones doing it. Systems like the What-if Machine, developed by Teresa Llano at Goldsmiths, University of London, and her colleagues, are being trained in the art of make-believe. The result could be machines that exhibit some of the most human-like AI yet seen. 'We are not in the business of making artificial humans, but of making computers that can better understand and interact with humans,' says Tony Veale at University College Dublin, Ireland, who is also involved in the What-if Machine project. 'We love stories, so we need our computers to adapt to this need.'

To do so, computers will need to see the world as we do – a giant leap for machine intelligence. No wonder many consider it to be one of the toughest challenges in AI. But we are getting there. The pay-off could not only be new ways of enjoying stories, but new ways of making sense of the world.

Telling tales is not easy. You have to pretend that things are other than they are. There are characters and motivations to untangle. Then there is a narrative to bind it all together. And, crucially, a good story needs to sit somewhere between dull and utterly implausible. 'Story generation pushes against some of the greatest problems in computer science,' says Michael Cook at Falmouth University, UK. It involves everything from choosing the best character to give an engaging viewpoint for the story, down to the nuts and bolts of generating individual sentences and natural-sounding language.

Early work on AI story generation in the 1970s focused on the problem of cause and effect in stringing together a narrative. One early, influential program was Tale-Spin, created in 1977 by James Meehan at the University of California, Irvine. The software generated stories involving animals, in the vein of Aesop's fables. A human user gave each character a goal and a library of plans through which to achieve it. If the user chose the correct mix of goals and plans, the characters behaved in such a way that a narrative emerged.

Character motivation

These systems were hit-and-miss, though. A key advance came with the introduction of overriding authorial goals that guided the characters' actions to a desired conclusion. Instead of acting independently, the characters could now be made to coordinate their actions to ensure they all ended up living happily ever after – or not, as the case may be. But too much coordination produces unsatisfying and unrealistic stories – and can give the impression that the characters are working together to bring about an author's goal.

Mark Riedl at the Georgia Institute of Technology, whose work on storytelling AI is supported by organizations as diverse as Disney and the US Defense Advanced Research Projects Agency, DARPA, tries to get around this problem by making his AI systems give characters motivations, avoiding the illusion of collusive behaviour.

Breaking news: bots get the scoop

At 6.28 a.m. on 17 March 2014 the *Los Angeles Times* published a story about an earthquake that had shaken California only three minutes earlier. The prose was informative but plain: 'A shallow magnitude 4.7 earthquake was reported Monday morning 5 miles from Westwood, California, according to the US Geological Survey. The tremblor occurred at 6.25 a.m. Pacific time at a depth of 5.0 miles.' The report appeared with the byline of Ken Schwencke, a journalist and programmer for the paper. But the credit ought have gone to Schwencke's computer, which wrote the story without human input.

Not that readers would necessarily have noticed. Earlier that month, Christer Clerwall at Karlstad University in Sweden asked 46 of his students to read one of two reports on an NFL American Football game and assess the quality and credibility of the article they read. Unknown to the students, one of the reports had been produced by a *Los Angeles Times* journalist, the other by a piece of software. Of the 27 people who read the computer-generated recap, nearly half believed that it had been written by a human.

Quick-fire, fact-based computer-generated journalistic reports are becoming increasingly prevalent because news stories, with their bald presentation of the facts, lend themselves to automatic generation. Creative writing is another matter. Whole stories can be told using the human connotations of everyday objects. Take Hemingway's six-word tragedy: 'Baby shoes for sale, never worn.' This goes far beyond news reports, demanding an intimate knowledge of the world.

Another problem with early systems was their reliance on hand-coded knowledge, which restricted the scope of their ostensible imagination. That is where the new wave of story-tellers is rapidly advancing. For example, one of Riedl's systems, called Scheherazade, learns by asking questions. When the AI recognizes that it doesn't know how to do something – such as how to make two characters meet at a restaurant – it posts a question to the Internet. Humans on crowdsourcing platforms like Amazon's Mechanical Turk then provide written examples of things that can happen in different scenarios, such as a first date or a bank robbery. The system learns about new situations from these examples, which it then weaves into stories.

Twists in the tale

There is more to a good story than just a blow-by-blow account of events, of course. Enjoyment often comes from an unexpected spin on the mundane. Understanding what properties objects have or what cultural meaning they might possess is crucial. It enables a storyteller to be inventive, creative and surprising. 'How do we teach an AI that a kettle can be used as a weapon, even though it's almost never used that way?' asks Cook.

Even if a computer were able to display a competent grasp of the myriad different systems and meanings that exist, there is still the challenge of how to make something up. One trick used by the What-if Machine, for example, is to invert things it knows about the world. Monkeys like bananas. What if they were afraid of them instead? Houses have doors. What if they didn't?

To judge whether an invention is new, though, an AI needs to compare it with what already exists. 'Let's say you've come up with the idea of a bear that is also a piece of furniture and you want to know if this revolutionary bear–chair hybrid is a novel idea,' says Cook. So you look at all the types of bear and furniture you know of and see if there's any overlap.

But an AI's judgement of novelty is only as good as the database of information from which it draws. 'An AI might think up an exciting new animal,' says Cook: 'A bird that can't fly!' If all the examples of birds it knows about can fly, this should rank highly on its novelty scale. But add penguins to the database and the idea is not so novel any more.

Again, learning from humans can help. But there's more to make-believe than bear–chairs and flightless birds. For Riedl, much of what makes a story interesting is whether or not the related events are unexpected. A story about a bank robbery in which everything happens as we might expect is unlikely to wow anyone. Narrative theorists often say that a story is only worth telling when it involves a breach of convention.

But it is not just a matter of breaking any old rule. Some breaches of convention are trivial, others nonsensical. Machines do not necessarily know when breaking a rule will be beneficial or harmful. An AI has to master the typical before it can start to reason about what is atypical.

One way to release computers from this trap is to teach them about metaphor. 'Metaphor is a cognitive lever that allows humans to project and magnify their knowledge of one domain into another,' says Veale. For example, saying 'life is a game' expands our concept of 'life'. To give computers similar leverage, Veale has created Metaphor Magnet, a program that learns from common metaphors in Google-cached texts. Using a thesaurus, it then unpacks the concepts in the metaphor to find new ones.

Three fiction factories

Are you sitting comfortably? The latest story-generation systems are learning to make things up by first figuring out how the human world works. Some canvas crowdsourcing platforms with questions about activities like first dates or bank jobs. Others mine online databases of concepts like ConceptNet to expand what they know about common objects and events before inverting them to invent fictions. Here are a few of the systems out there:

Scheherazade

Scheherazade, developed by Mark Riedl and colleagues at the Georgia Institute of Technology in Atlanta, can tell stories set in any world it can learn about via the Internet. Using crowdsourcing platforms like Amazon's Mechanical Turk, the system requests sample actions that can happen in various scenarios, such as a bank robbery. Here is part of a response from one crowd-worker:

> John opened the bank door.
> John went in the open bank door.
> John looked around the bank.
> Sally the bank teller noticed John.
> John approached Sally.
> Sally smiled at John.
> John said hello to Sally.
> John told Sally he had a gun.

Scheherazade strings together many sequences of actions like that one to get a sense of what actions are possible and the order in which they must appear. It then spins these snippets into an original story set in the given domain. Here is a composition based on the bank robbery scenario:

John took another deep breath as he wondered if this was really a good idea, and entered the bank. John stepped into line behind the last person and waited his turn. When the person before John had finished, John slowly walked up to Sally. The teller said, 'Hello, my name is Sally, how can I help you?' Sally got scared when John approached because he looked suspicious. John pulled out a handgun that was concealed in his jacket pocket. John wore a stern stare as he pointed the gun at Sally. Sally was very scared and screamed out of fear for her life. In a rough, coarse voice, John demanded the money. John threw the empty bag on to the counter. John watched as Sally loaded the bag and then grabbed it from her once she had filled it. Sally felt tears streaming down her face as she let out sorrowful sobs. John strode quickly from the bank and got into his car tossing the moneybag on the seat beside him. John slammed the truck door and, with tyres screaming, he pulled out of the parking space and drove away.

Here is another story generated in the same way, this time about a cinema date:

With sweaty palms and heart racing, John drove to Sally's house for their first date. Sally, her pretty white dress flowing in the wind, carefully entered John's car. John and Sally drove to the movie theatre. John and Sally parked the car in the parking lot. Wanting to feel prepared, John had already bought tickets to the movie in advance. A pale-faced usher stood before the door; John showed the tickets and the couple entered. Sally was thirsty so John hurried to buy drinks before the movie started. John and Sally found two good seats near the back. John sat down and raised the armrest so that he and Sally could snuggle. John paid more attention to Sally while the movie rolled and nervously sipped his drink. Finally working up the courage to do so, John extended his arm to embrace Sally. He was relieved and ecstatic to feel her move closer to him in response. Sally stood up to use the restroom during the movie, smiling coyly at John before that exit. John and Sally also held hands throughout the movie, even though John's hands were sweaty. John and Sally slowly got up from their seats. Still holding hands, John walked Sally back to his car through the maze of people all scurrying out of the theatre. The bright sunshine temporarily blinded John as he opened the doors and held them for Sally as they left the dark theatre and stepped back out into the street. John let go of Sally's hand and opened the passenger side door of his car for her but instead of entering the car, she stepped forward, embraced him, and gave him a large kiss. John drove Sally back to her home.

Flux Capacitor

Tony Veale at University College Dublin, Ireland, and colleagues have developed a system that generates 'character arcs' that can be used as the seeds of stories. The Flux Capacitor uses a metaphor generator to combine concepts into 'role transitions'. For example, two opposing concepts are picked – 'cute' and 'dreaded', say – and matched with plausible roles, such as 'cute clowns' and 'dreaded wizards'. Drawing on basic knowledge about the world, these roles are then strung together into a plausible character arc. Here are a few examples:

- What leads cute clowns to retire from circuses, to study necromancy and to become dreaded wizards?
- What pushes complaining protesters to tire of marching, to join faiths and to become uncomplaining devotees?
- What causes reputable journalists to be dismissed from news media, to embrace voyeurism and to become sleazy voyeurs?
- What pushes dumb actors to retire from acting, to attract flocks and to become godly preachers?
- What leads shabby beggars to regain homes, to go to medical school and to become tidy surgeons?

The Flux Capacitor uses the @MetaphorMagnet Twitter account to tweet its efforts and the team hopes to use the feedback from human followers to refine the system.

The What–if Machine

Teresa Llano at Goldsmiths, University of London and colleagues are building a system to generate Disney-like and Kafkaesque story ideas. The What–if Machine inverts the properties we commonly attach to concepts to create fictional scenarios:

Human characters

- What if there was a little man who forgot how to mow the lawn?
- What if there was a little lawyer who learned how to agree?
- What if there was a little baby who learned how to walk?
- What if there was a little person who learned how to melt?

Animal characters

- What if there was a little monkey who was afraid of bananas?
- What if there was a little dog who was afraid of love?
- What if there was a little dog who was afraid of bones?
- What if there was a little snake who was afraid of a live mouse?
- What if there was a little mole who couldn't find the hole?
- What if there was a little bee who couldn't find the honey?
- What if there was a little sheep who couldn't find the field?

Object characters

- What if there was a little wheel who lost his brake?
- What if there was a little book who lost his story?
- What if there was a little table who lost his chair?
- What if there was a little house who lost his door?
- What if there was a little bomb who forgot how to hurt a person?
- What if there was a little star who couldn't explode?
- What if there was a little pen who couldn't write?
- What if there was a little music who couldn't entertain?
- What if there was a little gun who couldn't kill?

Kafkaesque

- What if there was a woman who woke up in farm as a goat but could still talk?
- What if there was a man who woke up as a dog in a field, but could still use the telephone?

Surrealist

- What if there was a servant in a field who had the face of a turnip?
- What if there was a shepherd in a yard who had the face of a bell pepper?

Scenarios

- What if all poets stopped writing poems for fun and instead started drinking?
- What if there was an old dog, who couldn't run any more, which he used to do for fun, so decided instead to ride a horse?
- What if there was a robot that could only understand the concept of love by using theorems?
- What if there was a dancer who could only dance by using hands instead of feet?
- What if there was a lift that didn't have wires and it could rise as high as heaven?

From teacher to drug dealer

By analysing opposing concepts as well as related ones, Metaphor Magnet can also help deliver character arcs in a story. Take the *Breaking Bad* TV series, in which the main character evolves from a father and teacher into a drug dealer and criminal kingpin. The contrast between the character's roles at the start of the story and those at the end leads to a compelling narrative.

To generate similar arcs, Metaphor Magnet first identifies a pair of opposing concepts such as 'cute' and 'dreaded'. It then looks for roles that these concepts can be applied to – 'cute clowns' and 'dreaded wizards', say. Using a little knowledge of the world, these are then strung together into a plausible transition, providing the seed for a story: what leads cute clowns to retire from circuses, to study necromancy and to become dreaded wizards?

'A story about a CEO who becomes chairman of a company is very plausible, as CEOs are similar to chairpersons,' says Veale. 'But where is the tension? A story about an arrogant CEO who loses everything and becomes a bum? Now that's interesting.'

Working out what makes a good character arc is partly a matter of understanding suspense. Riedl's team has built a model that correlates suspense with the likelihood that a plan to get a character out of trouble will be successful. This lets Riedl's system evaluate the level of suspense in a plot.

The pieces of the puzzle are coming together. So what will we do with these systems? One practical use could be to generate stories that are too large for humans to maintain. Soon after Facebook acquired the virtual reality firm Oculus Rift, for example, it stated that it wants to build the first billion-user online role-playing game. Virtual worlds must be populated with interesting characters doing interesting things. If a world becomes big enough, it is no longer feasible for human games designers to hand-author characters, storylines and quests.

Future story-generating systems promise to be more than just fiction factories, though. Machines that tell stories will also understand how our world works. Our computers may surprise us, entertain us, provoke debate, reveal the possibilities of change, highlight contradictions and ironies, and generally prompt us to engage more on an intellectual level.

Riedl also believes that an AI that can master the basics of storytelling would be useful for factual analysis. AI investigative journalism could benefit from fictional-story generation by creating hypotheses for things that happened in the real world and then seeking additional facts to confirm or refute those hypotheses. For example, stories about what might have happened to a missing aircraft could guide a search.

Interview: Beyond the Turing Test

Mark Riedl is the director of the Entertainment Intelligence Lab at the Georgia Tech School of Interactive Computing in Atlanta. His work straddles artificial intelligence, virtual worlds and storytelling. He thinks the Turing Test is too easy – creativity should be the benchmark of human-like intelligence, he says – and so he has developed it into a new form, called the Lovelace 2.0 Test.

What are the elements of the Turing Test?

The Turing test was a thought experiment that suggested if someone can't tell the difference between a human and a computer when communicating with them using just text chat or something similar, then whatever they're chatting with must be intelligent. When Alan Turing wrote his seminal paper on the topic in 1950, he wasn't proposing that the test should actually be run. He was trying to convince people that it might be possible for computers to have human-like abilities, but he had a hard time defining what intelligence was.

Why did you think the test needed upgrading?

It has been beaten at least three times now by chatbots, which almost every AI researcher will tell you they don't think are very intelligent.

A 2001 test called the Lovelace Test tried to address this, right?

Yes. That test, named after the nineteenth-century mathematician Ada Lovelace, was based on the notion that, if you want to look at human-like capabilities in AI, you mustn't forget that humans create things, and that requires intelligence. So creativity became a proxy for intelligence. The researchers who developed that test proposed that an AI can be asked to create something – a story or poem, say – and the test would be passed only if the AI's programmer could not explain how it came up with its answer. The problem is that I'm not sure the test actually works, because it's very unlikely that the programmer couldn't work out how their AI created something.

How is your Lovelace 2.0 Test different?

In my test, we have a human judge sitting at a computer. They know they're interacting with an AI, and they give it a task with two components. First, they ask for a creative artefact such as a story, poem or picture. And secondly, they provide a criterion. For example: 'Tell me a story about a cat that saves the day'; or 'Draw me a picture of a man holding a penguin.'

Must the artefacts be aesthetically pleasing?

Not necessarily. I didn't want to conflate intelligence with skill: the average human can play Pictionary but can't produce a Picasso. So we shouldn't demand super-intelligence from our AIs.

What happens after the AI presents the artefact?

If the judge is satisfied with the result, they make another, more difficult, request. This goes on until the AI is judged to have failed a task, or until the judge is satisfied that it has demonstrated sufficient intelligence. The multiple rounds mean you get a score as opposed to a pass or fail.

And we can record a judge's various requests so that they can be tested against many different AIs.

So your test is more of an AI comparison tool?

Exactly. I'd hate to make a definitive prediction of what it will take for an AI to achieve human-like intelligence. That's a dangerous sort of thing to say.

The virtual virtuosos redefining creativity

Does it really take a human to produce a masterpiece? A few years ago, in a loft overlooking the rooftops of one of the buzzing artistic neighbourhoods of Paris, France, Simon Colton carefully unfurled one giant painting after another. One, called *The Dancing Salesman Problem* (see Figure 6.1), features colourful human figures dancing on a black background. The dancers are painted in long, flowing strokes, so they appear full of movement: they contort into beautiful poses and the bright colours bring the scene to life. The work could never be to everyone's taste, but you might have stopped to look at it in a gallery.

However, none of these paintings was the work of an ordinary artist. Nor were they by Colton, a computer scientist who was at Imperial College London at the time and is now at Falmouth University. Instead, they were created by a piece of software called the Painting Fool, which can seek artistic inspiration and, arguably, has a rudimentary imagination. It may have been designed by Colton, but its artwork is its own; the software did not base its composition on existing pictures.

The Painting Fool is one of a growing number of computers that, so their makers claim, possess creative talents. Classical music by an artificial composer has had audiences enraptured,

FIGURES 6.1 and 6.2 *The Dancing Salesman Problem* (top) and *Four Seasons*, two original works by the Painting Fool, an AI developed by Simon Colton.

and even tricked them into believing a human was behind the score. Artworks painted by a robot have sold for thousands of dollars and been hung in prestigious galleries. And software has been built which creates art that could not have been imagined by the programmer. 'It scares a lot of people,' says Geraint Wiggins, a computational creativity researcher at Goldsmiths,

University of London. 'They are worried it is taking away something special from what it means to be human.'

While some animals such as crows and monkeys have displayed traits that could be labelled as limited creativity, we are the only species to perform sophisticated creative acts regularly. If we can break this process down into computer code, where does that leave human creativity? 'This is a question at the very core of humanity,' says Wiggins.

To some extent, we are all familiar with computerized art. Software that is used to create or manipulate art is ubiquitous, but these are mere tools for a human artist. The question is: where does the work of a person stop and the creativity of the computer begin?

Consider one of the oldest machine artists, Aaron, a robot that has had paintings exhibited in London's Tate Modern and the San Francisco Museum of Modern Art. In some respects, then, Aaron passes some kind of creative Turing Test – its works are good enough to be exhibited alongside some of the best human art and people spend good money on them. Aaron can pick up a paintbrush with its robotic arm and paint on canvas on its own. Impressive, perhaps, but it can never break free from the tightly controlled rules it has been given by its programmer, the artist and founder of machine fine art Harold Cohen. Critics point out that Aaron is little more than a tool to realize Cohen's own creative ideas.

Different strokes

Colton is keen to make sure the Painting Fool has as much autonomy as possible. Although the software does not physically apply paint to canvas, it simulates many styles digitally, from collage to paint strokes. The Painting Fool needs only minimal direction and can come up with its own concepts by going online for source material. 'I don't even give it the notion of a person or a topic,' says Colton. 'It will wake up in the morning and look at the newspaper headlines.' The software runs its own web searches and trawls through social media websites such as Twitter and Flickr.

The idea is that this approach will let it produce art that is meaningful to the audience, because it is essentially drawing on the human experience as we act, feel and argue on the web. In 2009 Colton and graduate student Anna Krzeczkowska asked the Painting Fool to produce its own interpretation of the war in Afghanistan, based on a news story. The result is a striking juxtaposition of Afghan citizens, explosions and war graves. 'This piece struck a chord with me, and shows the potential for the software to add poignancy and intentionality to its paintings,' says Colton.

The Painting Fool can also create pictures from scratch. One of its original works, part of a series that Colton has called *Four Seasons* (see Figure 6.2), depicts fuzzy panels of simple landscapes. It is hard to judge how good it is without applying double standards towards software-produced and human-produced art, however. Colton argues that we should remember that the Painting Fool painted the landscapes without referring to a photo. 'If a child painted a new scene from its head, you'd say it has a certain level of imagination, even if it's just a little bit,' he says. 'The same should be true of a machine.'

Software bugs can also lead to unexpected results. A few works in the Painting Fool's oeuvre were happy accidents. Some paintings of a chair came out black and white thanks to a glitch, for example. It gives the work an eerie, ghostlike quality. Human artists like Ellsworth Kelly are lauded for limiting their colour palate – so why should computers be any different?

New kinds of play from happy accidents

Michael Cook at Falmouth University – who is a colleague of Simon Colton – has developed an AI called Angelina that can design its own video games. Cook sees games as the perfect medium for exploring computer creativity because they draw on multiple disciplines at once, from sound and visual design to picking rules that lead to an engaging experience for the player.

Like Colton, Cook has also found that software bugs can lead to his system making innovative leaps. Angelina makes games using several different techniques – including reading the news online and incorporating themes it finds into its games, as the Painting Fool does with paintings. Angelina can also use the code of existing games as a starting point and refine the features it finds into something new.

The ability to pick and choose design ingredients was a big advance, says Cook. Previously, the system came up with game mechanics by putting together rules it was given. 'It would slot them together in new ways like a jigsaw, but I was never very happy with it,' says Cook. 'After all, it needed me to hand it the jigsaw pieces.'

Angelina finds and tests game possibilities – like reversing gravity, high jumping and teleportation – on its own. Cook starts things off by providing a game level that cannot be solved, such as one with a wall between the start and the exit. Angelina then redesigns the level in an iterative process, using ideas it finds in existing games – making changes, testing them, and making further tweaks until the level works. 'It's closer to what a human does when they program,' says Cook.

Even more cunningly, it has found bugs in Cook's code and taken advantage of them to invent new game levels. In one case, the game code wrongly let a player teleport inside a wall and still allow the character to jump. So Angelina invented a wall-jumping technique, where the player could climb up a vertical wall by repeatedly teleporting and jumping. 'This was why I felt it's so important to create a system that was independent of me,' says Cook.

In another example, Angelina found code that could be used to make the player bouncy, something that Cook had been unaware of. 'I've only seen a few games that use bouncing in that way,' Cook says. 'You can't even guarantee that professional developers will think of these things.'

Angelina is also the first non-human to take part in a game jam – an informal competitive event where people get to together to make a new game in just a few days. To save time, Angelina was given game rules that were a variation of a pre-coded template. But the rest, including the aesthetic choices, was the work of the AI. The game it came up with was set in a place with blood-red walls and unsettling music. The atmosphere this creates is striking. Other players judged Angelina's entry without knowing it was the work of an AI and described the result as 'creepy' and having 'a weird little unsettling vibe' – positives from the point of view of creating an engaging experience.

A controversial composer

Researchers like Colton do not believe it is right to compare machine creativity directly with our own, when we have had millennia to develop our skills. Others, though, are fascinated by the prospect that a computer might create something as original, emotional and subtle as our best artists. So far, only one has come close.

One day in 1981 David Cope was suffering from 'composer's block'. He had been commissioned to write an opera but was struggling to come up with the goods. If only a computer could understand his style, he thought, and help him write new material. That idea was the starting point for what was to become one of the most controversial pieces of creative software to date. Cope came up with a program called Experiments in Musical Intelligence, or EMI. He fed in musical scores and out popped new material in the composer's style. Not only did EMI create compositions in his style but also those of the most revered classical composers, including Bach and Mozart.

To an untrained ear, it sounds like any other classical music – and at times rich and emotional. Audiences who have heard the music have been moved to tears, and EMI even fooled classical music experts into thinking they were hearing genuine Bach. If ever there were a successful Turing Test for computational creativity, this had to be it.

Not everyone was impressed, however. Some critics, such as Wiggins, have blasted Cope's work as pseudoscience, saying his explanation of how the software works is 'smoke and mirrors' leaving others unable to reproduce the results. Douglas Hofstadter, at Indiana University, Bloomington, says Cope merely scratches at the surface of creativity, using superficial elements of an artist's work to create replicas, which still rely on the original artist's creative impulses.

Nonetheless, for others EMI's ability to mimic Bach or Chopin has serious implications. If it is so easy to break down the style of some of the world's most original composers into computer code, that means some of the best human artists are more machine-like than we would like to think. Indeed, when audiences found out the truth about EMI they were often outraged – one music lover allegedly told Cope he had 'killed music' and tried to punch him. Amid such controversy, in 2004 Cope decided that EMI's time was up, and he destroyed its vital databases.

Why did so many people love the music yet recoil when they discovered what composed it? A study by David Moffat, a computer scientist at Glasgow Caledonian University in the UK, provides a clue. He asked both expert musicians and non-experts to assess the creative worth of six compositions. The participants were not told beforehand whether the tunes were composed by humans or computers, but were asked to guess, and then rate how much they liked each one. Perhaps unsurprisingly, people who thought the composer was a computer tended to dislike that piece more than those who believed it was human. This was true even among experts, who you might think would be more objective in their analysis of musical quality.

Where does this prejudice come from? Psychologist Paul Bloom of Yale University has a suggestion: he reckons part of the pleasure we get from art comes from our perception of the creative process behind it. This can give it what Bloom calls an 'irresistible essence'. This idea explains why a painting loses its value if it is exposed as a fake, even though we might have loved it when we thought it was an original. Indeed, experiments by psychologist Justin Kruger of New York University have shown that people's enjoyment of an artwork increases if they think more time and effort were needed to create it.

Similarly, Colton thinks that when people experience art, they engage in a discourse with the artist. We wonder what the artist might have been thinking, or ponder what they are trying to tell us. With computers producing art, this speculation is cut short – there's nothing to explore. But as the software becomes increasingly complex, finding those greater depths in the art may become possible. That is why Colton asks the Painting Fool to tap into online social networks for its inspiration: in the hope that this way it will choose themes that will already mean something to us.

Unconscious creativity

Douglas Hofstadter thinks that the more complex machines become, the more easily we will accept their art – especially if they can interact more with the physical world. If robots bumped into things and had goals, successes and failures, then that might be enough. 'They would be sort of pathetic and laughable and once in a while heroic,' he says. 'I don't think people would be uncomfortable with creatures like that writing an essay or composing a piece of music or painting a picture.' Yet the fact that machines now lack this kind of self-awareness is perhaps the most irksome element of com-putational creativity. How can you be creative without even being conscious?

Surprisingly, consciousness might not be as crucial to creativ-ity as we like to think. Our brains work creatively even when we aren't consciously thinking about it, says Arne Dietrich, a neuroscientist at the American University of Beirut in Lebanon. Just think back to a time when the solution to a problem you had forgotten about just popped into your head. There are sev-eral different types of creativity – some of them conscious, some of them unconscious. Creativity can happen when you delib-erately try to create something or it can happen in your sleep.

In any case, Dietrich believes that the creative brain might work much like software. Neuroscientists suspect that creativ-ity is essentially about discovery rather than anything mystical – driven by a mechanical process in the brain that generates possible solutions and then eliminates them systematically. He believes our tendency to dismiss computational creativity as inferior to our own comes from an ingrained dualism in human culture. 'We are over-valuing ourselves and underestimating them,' he says.

As a neuroscientist, Dietrich says he tackles the brain as a machine – and does not see machine creativity as different. Considered in this way, the idea that the human brain has a unique claim to creative talents seems a limited perspective. Will others accept that idea? The trick, says Colton, is to stop trying to compare computer artists to human ones. If we can embrace computer creativity for what it is and stop trying to make it look human, not only will computers teach us new things about our own creative talents, but they might become creative in ways that we cannot begin to imagine. They are creating a whole new form of art with the potential to delight, challenge and surprise us.

The muse in the machine

Will handing over the keys to creativity take something away from being human? For Michael Cook, a future in which an AI is able to create works of art will not rob us of anything – quite the opposite, in fact. He thinks that AI has a major role to play in democratizing creativity and lowering barriers of entry for people. If an AI can write a story or paint, then it can critique a story. And that means it can act as an assistant to people who want to create something themselves but don't know where to start, or who struggle with some aspect of it. Cook points to spellcheckers and tools in photo-editing software as examples of where computers could offer better help. 'Right now, that's a lousy level of involvement,' he says. 'We want to build software that can be a mentor, a muse and an audience all at once.'

Interview: How do you teach a computer to create?

Simon Colton is Professor of Digital Games Technology at Falmouth University, UK. He works on software that behaves in ways that would be deemed creative if seen in humans – such as painting the image of him in Figure 6.3. Software can already be artistic or make mathematical finds, but for it to excel we have to give it the right skill set, he says.

You designed software called HR to make its own discoveries. Has it had much success?

One thing HR came up with was a classification of mathematical structures known as Latin squares. Like a Suduko puzzle, these are grids of symbols where each row and column contains every symbol. HR produced some of the first algebraic classifications of these structures. A version of HR also independently came up with Goldblach's conjecture – that every even integer greater than 2 can be expressed as the sum of two primes.

Are mathematicians interested in using the system?

We found that mathematicians like software to do the boring menial work – the massive calculations and trivial proofs they know are true. But creative things like inventing concepts and spotting conjectures they like to do themselves. I once sent Herbert Simon, the Nobel prizewinning economist and computer scientist, an email about a conjecture that HR had proved. He later told me that he hadn't read to the end because he wanted to solve the puzzle himself. His wife said she had to call him to bed.

How do you make software discover things?

You give it data that you want to find something out about, but rather than looking for known unknowns – as with machine learning, where you know what you're looking for but not what it looks like – it tries to find unknown unknowns. We want software to surprise us, to do things we don't expect. So we teach it how to do general things rather than specifics. That contradicts most of what we do in computer science, which is to make sure software does exactly what you want. It takes a lot of effort for people to get their heads round it.

Can computers make breakthroughs?

I think we will only see computers making true discoveries when software can program itself. The latest version of HR is specifically designed to write its own code. But it's a challenge; it turns out that writing software is one of the most difficult things that people do. And, ultimately, there are mathematical concepts that you can't turn into code, especially ones dealing with infinity.

Another program of yours, the Painting Fool, creates portraits. How do people respond to this type of creativity?

Mathematicians will accept a computer as being creative if it produces great results over and over again. But in the art world, people take more convincing. When you buy a painting, you buy it for many reasons, only one of which is that it will look good with your sofa. When you like a painting, you're celebrating the humanity that went into it. How can we get software to fit in with that?

I don't want to do Turing Tests where we try to confuse people about who or what is doing something. We want

FIGURE 6.3 A portrait of Simon Colton by the Painting Fool

people to relate to what the software does on its own terms. But computers won't replace people in the creative industries because we will always pay for humanity – for blood, sweat and tears.

AI's artistic development

1973

Aaron paints abstract images that have been exhibited in London's Tate Modern and the San Francisco Museum of Modern Art.

Novel Writer, the first automatic story-generation system, creates murder stories set during a weekend party. The stories emerge via the simulated actions of a given set of characters.

1977

Tale-Spin generates stories about woodland animals in the vein of Aesop's fables. A human user gives characters goals and sets of actions with which to achieve them. The stories emerge from the characters' simulated interactions.

2010

An AI called Angelina starts creating its own video games.

David Cope's Emmy evolves into a later system called Emily Howell, which releases its debut album *From Darkness, Light.*

Francisco Vico and his colleagues at the University of Malaga in Spain develop Ianus, a computer that used an evolutionary technique called melomics (a portmanteau of 'genomics of music') to compose music without human guidance. Ianus's music is performed at Alan Turing's 2012 centenary concert and recorded by the London Symphony Orchestra.

2006

The Painting Fool paints portraits in different styles, depending on the AI's 'mood'.

2013

Scheherazade, developed by Mark Riedl and colleagues at the Georgia Institute of Technology in Atlanta, can tell stories set in any domain it can learn about via the Internet.

2014

IBM's Watson starts creating its own recipes.

The What-if Machine starts to generate Disney-like and Kafkaesqe story ideas.

The Flux Capacitor uses an underlying metaphor generator to combine concepts into the seeds of stories.

Ianus releases an album of pop music called *Omusic.*

1981

Author, a system designed to model an author's mind, is the first to introduce authorial goals on top of those of a story's characters. The system is thus able to generate stories that work, for instance, towards a happy ending.

1983

Universes generates stories for a series of soap opera episodes with many characters and overlapping storylines without endings.

2004

Frédéric Fol Leymarie and Patrick Tresset at Goldsmiths, University of London, create Aikon, a robot that draws portraits in Tresset's style by mimicking the way his wrist flexes and the pressure he applies to his pen.

1987

Composer and computer scientist David Cope makes Experiments in Musical Intelligence, or EMI ('Emmy'), software that automatically composes music in the styles of different human composers such as Beethoven, Chopin and Vivaldi by learning from human feedback.

7
The real risks of AI

Why fears of an apocalypse are overblown

The seemingly relentless advance of AI has been followed by a wave of public anxiety. Several high-profile figures have spoken out about the existential risk AI could pose to humanity. But fears of AI one day triggering an apocalypse are completely overblown. That is not to say that there are no risks, however. There are many ways in which AI could change our world for the worse.

Forget Skynet: the societal effects of AI

Hypothetical world-ending artificial intelligence makes headlines, but the hype ignores what is happening right under our noses. Some have fearfully predicted that these intelligent machines will dispense with useless humans, while others see a utopian future filled with endless leisure.

Focusing on these equally unlikely outcomes has distracted the conversation from the very real societal effects already brought about by the increasing pace of technological change. For 100,000 years we relied on the hard labour of small bands of hunter-gatherers. A scant 200 years ago we moved to an industrial society that shifted most manual labour to machines. And then, just one generation ago, we made the transition into the digital age. Today much of what we manufacture is information, not physical objects – bits, not atoms. Computers are ubiquitous tools and much of our manual labour has been replaced by calculations.

This rapid shift has stoked no end of paranoia, but a reality check is needed. Cherished qualities like creativity and invention may well be outsourced to AI in the coming years. But we should not feel threatened by this: we should feel exhilarated at the new things we can do with their help, just as the digital tools we use today have enhanced and diversified the ways in which we communicate and create.

It is not that the boom in AI gives us nothing to worry about. However, it is not the technology itself that should concern us so much as how we design and use it. AI is already conferring great power on those who control it. A large number of breakthroughs are made by corporations or other for-profit organizations. Whose interests do they serve? In this, as elsewhere, the answers lie with us. AI can't strip us of our jobs, our dignity or our human rights. Only other humans can do that.

Data streams

Intelligent machines need to collect data – often, personal data – in order to work. This simple fact potentially turns them into surveillance devices: they know our location, our browsing history and our social networks. Can we decide who has access, what use can be made of the data, or whether the data gets deleted for ever? If the answer is no, then we don't have control.

Another concern is persuasion. The business model of many AI companies is advertising, which means getting people to click on specific links. Research on how to steer users is well under way. The more the machines know about us, the better the job they can do of nudging us. Predictive interfaces might even induce addiction in vulnerable users, by actively rewarding them with the juiciest content that the web has to offer. This is something that needs to be carefully studied.

AI has come a long way from its early days in academic laboratories. It is now being integrated everywhere into our lives. Often we no longer even think of it as AI once it is deployed, behind the scenes. But we might want to resist the temptation to introduce AI into as many domains as possible, at least before the cultural and legal framework evolves. Widespread adoption of AI brings remarkable opportunities but also potential risks. These are not existential risks to our species but, rather, a possible erosion of our privacy and autonomy.

Seeking employment

The British economist John Maynard Keynes (1883–1946) always assumed that robots would take our jobs. In 1930 he wrote that it was all down to 'our means of economizing the use of labour outrunning the pace at which we can find new uses for labour'. And that was no bad thing. Our working week would shrink to 15 hours by 2030, he reckoned, with the rest of our time spent trying to live 'wisely, agreeably and well'.

It has not happened like that – indeed, if anything, many of us are working more than we used to. Advanced economies that have seen large numbers of manual workers displaced by automation have generally found employment for them elsewhere, for example in service jobs. The question is whether that can continue, now that AI is turning its hand to all manner of tasks beyond the mundane and repetitive.

Fear of machines taking jobs dates back at least as far as the Luddites, a group of British weavers who went on a mill-burning rampage in 1811 when power looms made them redundant. Two centuries on, many of us could face the same predicament. In 2013 Carl Frey and Michael Osborne of the Oxford Martin Programme on the Impacts of Future Technology at the University of Oxford looked at 702 types of work and ranked them according to how easy it would be to automate them. They found that within two decades, machines could feasibly do just under half of all jobs in the United States.

The list included jobs such as telemarketers and library technicians. Not far behind were less obviously susceptible jobs, including models, cooks and construction workers, threatened respectively by digital avatars, robochefs and prefabricated buildings made in robot factories. The least vulnerable included mental health workers, teachers of young children, clergy and choreographers. In general, jobs that fared better required strong social interaction, original thinking and creative ability, or very specific fine motor skills of the sort demonstrated by dentists and surgeons.

Which jobs will go next?

Artificial intelligence is close to taking on a number of human jobs. Here are three that could be the next to go human-free:

- **Taxi drivers:** Uber, Google and established car companies are all pouring money into machine vision and they also control research. Legal and ethical issues will hold back the process but, once it starts, human drivers are likely to become obsolete.
- **Transcribers:** Every day, hospitals all over the world fire off audio files to professional transcribers who understand the medical jargon doctors use. They transcribe the tape and send it back to the hospital as text. Other industries also rely on transcription. Slowly but surely, machine transcription is starting to catch up, much of it driven by data on the human voice gathered in call centres.
- **Financial analysts:** Kensho, based in Cambridge, Massachusetts, is using AI to instantly answer financial questions, which may take human analysts hours or even days to answer. By digging into financial databases, the start-up can answer questions like: 'Which stocks perform best in the days after a bank fails?' Journalists at NBC can already use Kensho to answer questions about breaking news, replacing a human researcher.

An automated workforce

AI has already taken on a range of jobs, from organizing nightly maintenance on Hong Kong's subway system to helping out with subtle legal research, as does ROSS, an AI assistant built on IBM's Watson computer. In the next few years AI looks set to cause at least short-term turbulence in the labour market.

Between 2012 and 2015 UK telecoms firm O2 replaced 150 workers with a single piece of software. A large portion of O2's customer service is now automatic, says Wayne Butterfield, who works on improving O2's operations. Sim swaps, porting mobile numbers, migrating from prepaid to a contract, unlocking a phone from O2 – all are now automated.

Humans used to manually move data between the relevant systems to complete these tasks, copying a phone number from one database to another, for example. The user still has to call up and speak to a human, but now an AI does the actual work.

O2's AI learned on the job. It watched humans do simple, repetitive database tasks and then went off to work on its own. 'They do exactly what a human does,' says Jason Kingdon, chairman of Blue Prism, the start-up that developed O2's artificial workers. 'If you watch one of these things working, it looks a bit mad. You see it typing. Screens pop up, you see it cutting and pasting.'

Barclays, one of the world's largest banks, has also employed AI to do back-office work. It used Blue Prism to deal with the torrent of demands that poured in from its customers after UK regulators demanded that it pay back billions of pounds of mis-sold insurance. It would have been expensive to rely entirely on human labour to field the sudden flood of requests. Having software agents that could take some of the simpler claims meant Barclays could employ fewer people.

Kingdon does not shy away from the implications of his work: 'This is aimed at being a replacement for a human, an automated person who knows how to do a task in much the same way that a colleague would.'

Another problem with new kinds of automation like Blue Prism and ROSS is that they perform the kinds of jobs that are the first rung on corporate ladders, which could result in deepening inequality as fewer opportunities are made available to new job seekers.

Robot co-workers

Others find fears of widespread unemployment overblown. A recent working paper for the rich-world OECD club suggests that AI will not be able to do all the tasks associated with all these jobs – particularly the parts that require human interaction – and only about 9 per cent of jobs are fully automatable. What's more, past experience shows that jobs tend to evolve around automation.

According to this more Keynesian view, technological progress will continue to improve our lives. The most successful innovations are those that complement rather than usurp us. Witness the prominence of 'cobots' at the annual automation expo in Chicago in 2016, for example. Such robots are designed to work alongside people, making their work safer and easier, not replacing them. They could help us solve problems, communicate widely, or create art, music and literature. The weight of expert opinion is behind this view. In 2014 the Pew Research Center, a US think tank, asked 1,896 experts whether they thought that by 2025 technology will have destroyed more jobs than it creates. The optimists outnumbered the pessimists.

That is not to say that AI will shake things up. Even if it changes the nature of work, rather than replacing workers the impact on society could be big. The gig economy, pioneered by firms such as Uber, adds flexibility to the labour force and convenience to their customers via algorithmic management – but at a cost to workers' rights and conditions. And AI could accelerate that trend.

FIGURE 7.1 Repetitive manual jobs have already been hit by automation, but AI will have a big effect on a wide range of professions.

This matters: our work is integral to our identities, and preserving the dignity of labour should be central to our society. We should strive to ensure that AI is used to upskill workers rather than paring their jobs down to tedious piecework: dehumanizing workers is a poor use of the technology (see Figure 7.1).

Tackling that is a social and political issue rather than a technological one. AI may force changes on our economic system – witness the discussions over the introduction of a basic universal income for all. But change should be people-centric, not led by AI-driven efficiency that enriches a few to the detriment of the many. Ultimately, of course, we are in charge of our own destiny. Given the benefits of work for our health and wellbeing, we may opt to protect fulfilling, rewarding work. There will be inequities and disruptions, but that has been the case for hundreds of years.

Silicon Valley's hot new job: a bot's assistant

Have we been thinking too much about the jobs AI could take away, rather than those it could create? Services like Facebook's digital personal assistant, M, reveal at least one new role some of us may be taking on in future.

M is an AI-backed digital assistant built into Facebook messenger. It can book your next hotel or flight, recommend a restaurant and reserve a table, purchase items for delivery or send news updates and reminders. What's the high-tech secret sauce that makes M tick? It's humans – or, in Facebook parlance, AI trainers. Ask M to recommend a local restaurant with good Pad Thai and an AI trainer will review its suggestions before they are sent back to you. Tell it to reserve a table for two and it may be an AI trainer who actually picks up the phone. Everything that M says is observed, validated or tweaked by a hired human being. 'We've invented a new kind of job,' says Ari Entin, a Facebook spokesman.

Facebook is not the only tech company to think of using humans as a hack. Clara Labs, a start-up in San Francisco, builds a virtual assistant that you can email to help set appointments in your calendar. Clara is an AI, but when you email her, you are also unwittingly conversing with a number of humans who check over her work.

Interactions, based in Franklin, Massachusetts, is another company building 'digital conversational assistants'. These entities handle the customer service hotlines for big corporations, such as US health insurance company Humana and Texas utilities company TXU Energy. Interactions calls its human helpers 'intent analysts'. When the automated assistant comes up against a string of words it can't quite understand, it sends them to an intent analyst for interpretation. The human listens in, tells the software what to do next, and the caller's conversation is back on track.

This is an easy way to fake great AI – but companies like Facebook, Clara Labs and Interactions are not just involved in an elaborate ruse. It shows that engineers have learned a lot about how valuable humans can be. Automated helplines have a bad reputation. One online service called GetHuman.com gives people special tips and tricks for different corporate contact numbers that are meant to ensure that you speak to a real person.

Why not just stick with people in the first place? Plenty of Silicon Valley assistant apps have gone down that route. One called Magic lets you text requests to a team of operators who can grant them, whether it involves food delivery, reservations or even medical marijuana. Invisible Girlfriend, based in St Louis, Missouri, lets you invent a fake sweetheart and text with her for a monthly fee. In this case, 'she' is a crowd of human workers who take turns crafting flirty or sentimental responses.

Since people can be troublesome and expensive, an AI backed by human assistants combines the flexibility and creativity of the human brain with the tireless speed – and frugality – of automation. The benefit for companies is obvious. But what is it like to be a bot's assistant?

The offices at Interactions look like call centres, says vice president Phil Gray, but they don't sound like them. The clamour of conversation has been replaced by the non-stop tapping of keyboards. 'Some people compare it to video game playing,' adds Jane Price, another vice president at the company. On Glassdoor, a website where employees can leave anonymized reviews of their workplace, comments about Interactions are mixed. Some praise the casual atmosphere and flexible schedules, and say they enjoy the fast pace of the work. Others were numbed by its repetitive nature. 'You will feel like you're turning into a zombie sometimes,' wrote one former analyst.

Whatever you call these brave new workers – AI trainers, intent analysts – they serve a dual purpose. For now, they function as an AI's backup, filling in gaps when the software encounters a problem it cannot yet handle. But they are also there to teach the AI not to make those mistakes again. Each instance of coaching is added to a growing library of training data, which machine-learning algorithms can draw on to handle unfamiliar tasks in future.

Does this mean that at some point the AI will be fully trained, leaving the human trainers out of a job? Alex Lebrun, who is in charge of M at Facebook, says it is not that simple. 'We will always need the trainers,' he says. 'The more we learn, the more there is to learn. It is never-ending learning.'

Office spies

Taking jobs is not the only way that AI could be a problem, however. Companies are increasingly using technology to monitor employees in the workplace, with AI making it possible to track individuals' behaviour in great detail. Start to slack off or show signs of going rogue, and an algorithm could tattle to your boss.

One firm offering such services is London start-up StatusToday. The company was included in a cybersecurity accelerator run by British intelligence agency GCHQ, which offers technical expertise and helps to secure investment. StatusToday's AI platform relies on a regular supply of employee metadata, including everything from the files you access and how often you look at them to when you use a key card at a company door.

The AI uses this metadata to build up a picture of how companies, departments and individual employees normally function, and then to flag anomalies in people's behaviour in real time. The idea is that it could detect when someone might pose a security risk by stepping outside their usual behavioural patterns. 'All of this gives us a fingerprint of a user, so if we think the fingerprint doesn't match, we raise an alert,' says Mircea Dumitrescu, the company's chief technology officer.

The system could point out if an employee starts copying large numbers of files they don't normally view, for example. Are they just going about their job or are they stealing confidential information? It also aims to catch employee activity that could lead to a security breach, like responding to a phishing email or opening a malware-carrying attachment. 'We're not monitoring if your computer has a virus,' says Dumitrescu. 'We're monitoring human behaviour.'

But catching the odd security breach in this way means monitoring everyone. Some companies already keep employee metadata for retrospective analysis if something goes wrong. Insurance firm Hiscox recently started using the StatusToday platform and immediately detected activity on an account from an employee who had left the company months earlier.

In addition to flagging potential cybersecurity alerts, the AI can be used to track employee productivity. Dumitrescu cites the example of Yahoo controversially banning staff from working from home on the basis that this reduced the 'speed and quality' of work across the board. 'We can actually quantify if this is true for individual employees,' he says. 'Whether they should be allowed to work from home can then be based on data.'

Is your photo in a police database?

If you live in the United States, there is a 50:50 chance that you are in a police face-recognition database, according to a 2016 report from the Georgetown University Law Center in Washington DC. The report suggests that around a quarter of all police departments in the United States have access to face-recognition technology.

That police are using face-recognition technology is not a problem in itself. In a world with a camera in every pocket, they would be foolish not to. But face recognition can be used far more broadly than fingerprint recognition, which means it carries a higher risk of tagging innocent people. 'It's uncharted and frankly dangerous territory,' said Alvaro Bedoya, who led the Georgetown study, in a statement when the report was published.

Fingerprints are difficult to work with. Prints from known criminals can only be gathered in controlled environments at police stations, and dusting for prints is so time consuming that it is only done at relevant crime scenes. This narrows down the number of people in the sights of any one investigation. It is much easier to build huge databases of identified photographs. The majority of the 117 million faces in the police data sets come from state driving licences and ID cards. And when trying to solve a crime, gathering faces is as easy as pointing a camera at the street. People attending protests, visiting their church or just walking by can all have their faces 'dusted' without ever knowing it.

Some face-recognition software has performed as well as humans in controlled tests, but systems dealing with grainy CCTV images fare far worse. If software starts turning up more false matches than human investigators can check, it will make it harder to fight crime. There are also few regulations on how the police use this technology or how much weight they give to its results. Without guidance, officers may overvalue the output of a face-recognition system and favour evidence that matches its results.

Such systems are also likely to be biased against black people. Since black people are arrested more often than white people, black faces are over-represented in the mugshot databases. This means that innocent black people are more likely to be linked to a crime by face recognition than innocent white people. What is more, studies have shown that commercial face-recognition software is less accurate at analysing the faces of black people, women and children than those of white men. Not only is the software more likely to point the finger at black people but it is also more likely to be wrong.

None of the four main companies selling face-recognition technology – Cognitec, NEC, 3M Cogent and Morpho – is open about how their software works or what data sets they use to train it. Not even the FBI knows exactly what it is doing. In 2016 the US Government Accountability Office (GAO) released a report on the FBI's face-recognition programme, stating that the agency had not checked to see how often errors occurred. By conducting better tests, the GAO said, the FBI could be surer that its system 'provides leads that help enhance, rather than hinder, criminal investigations'.

Like all forensic techniques, face recognition has the power to catch criminals police might otherwise miss. But to do so, its results must be transparent and reliable – otherwise you might as well just pick someone out of a crowd.

No one in control

Software is now involved in making decisions that can change people's lives. There are automated systems that help decide who gets a bank loan, who gets a job, who counts as a citizen and who should be considered for parole. Yet we only need to look at what happens when price-setting algorithms run amok to see the potential risks. Who will step in when the machines get out of hand?

'Amazon is all kinds of broken.' If you caught that tweet on 12 December 2014, and were quick, you might have grabbed some exceptional bargains. For one hour only, Amazon was selling an odd mix of items – phones, video games, fancy-dress costumes, mattresses – for one penny.

The surprise price drop cost sellers dearly. Goods usually marked £100 went for a 99.99 per cent discount. Hundreds of customers leapt at the chance, often buying in bulk. Even though Amazon reacted quickly and cancelled many orders, they were unable to recall those that their automated system had already dispatched from warehouses. Once set in motion, the process was hard to stop. Thanks to a software glitch, a handful of independent traders using Amazon's Marketplace lost stock worth tens of thousands of dollars. Some faced bankruptcy.

Automated processes are no longer simply tools at our disposal: they often make the decisions themselves. Many are proprietary and all are complex, pushing them beyond public scrutiny. How can we be sure they are playing fair? A new wave of algorithm auditors is on the case, intent on pulling back the curtain on the hidden workings and hunting for undue bias or discrimination.

It can be hard to predict how software will behave with real-world data, once released into the wild. When Microsoft launched its chatbot Tay in 2016 it was just a matter of hours before it was parroting racist comments on Twitter, forcing Microsoft to pull the plug. The scope of software's influence is also often unclear. Some people swear that they've seen the price of flights on one website jump after checking out a rival site, for example. Others think that this is an urban myth for our times. Such debates highlight the shadowy nature of today's systems.

Crash damage

The potential effects could be devastating, however. Some think hidden algorithms played a part in the 2008 sub-prime mortgage crash. Between 2000 and 2007 US lenders like Countrywide Home Loans and DeepGreen doled out home loans at an unprecedented rate via automated online applications.

The problem was granting so many high-risk loans without human oversight. Americans from minority groups suffered most in the resulting crash. Automated processes crunched through vast amounts of data to identify high-risk borrowers – who are charged higher interest rates – and targeted them to sell mortgages. 'Those borrowers turned out to be disproportionately African-American and Latino,' says Seeta Gangadharan of the Open Technology Institute, a public policy think tank based in Washington DC. 'Algorithms played a role in that process.'

The exact degree to which algorithms were to blame remains unclear. But banks like Wells Fargo and Bank of America settled with several cities, including Baltimore, Chicago, Los Angeles and Philadelphia, for hundreds of millions of dollars over claims that their sub-prime lending had disproportionately affected minorities. Although the decision-making process that big banks used to target and sell sub-prime loans may not have been new in itself, the reach and speed of those decisions when algorithms were the driving force was new.

Automated systems are replacing human discretion in ever more important decisions. In 2012 the US State Department started using an algorithm to randomly select the winners of the green card lottery. The system was prone to bugs, however: it awarded visas only to people who applied on the first day, says Josh Kroll, a Princeton University computer scientist who has investigated the event. Those visas were rescinded, but it is yet another illustration of how hidden algorithms can have a life-changing effect.

Modelled citizens

In a similar example, the documents leaked by Edward Snowden revealed that the National Security Agency uses algorithms to decide whether a person is a US citizen. According to US law, only non-citizens can have their communications monitored without a warrant. In the absence of information about an individual's birthplace or parents' citizenship, the NSA algorithms use other criteria. Is this person in contact with foreigners? Do they appear to have accessed the Internet from a foreign country?

Depending on what you do online, your citizenship might change overnight. 'One day you might be a citizen, another you might be a foreigner,' says John Cheney-Lippold at the University of Michigan in Ann Arbor. 'It's a categorical assessment based on an interpretation of your data, not your passport or your birth certificate.'

Some of the most egregious examples involve biases baked directly into code, hidden under a veneer of mathematical precision. Take prison sentencing. Judges and lawyers in certain US states can use an online tool to make an 'Automated Sentencing Application'. The system calculates incarceration costs for defendants and weighs them against the likelihood that the defendant will reoffend, based on prior criminal history and behavioural and demographic factors. Yet proxies like address, income and education level make it almost impossible to avoid racial bias. In 2016 investigative website ProPublica analysed one program in Broward County, Florida, and found that it falsely labelled black people as future reoffenders almost twice as often as white people over a period of two years.

In many of these examples, the problem is not the algorithms themselves but the fact that they over-amplify an existing bias in the data. What should we do about it?

Higher standards

Christo Wilson, at Northeastern University in Boston, thinks that large technology companies like Google and Facebook ought to be considered public services that huge numbers of people rely on. 'Given that they have a billion eyeballs, I think they have a responsibility to hold themselves to a higher standard,' he says.

Wilson thinks that automated systems might be made more trustworthy if users can control exactly how their results are personalized – such as leaving gender out of the equation or ignoring income bracket and address. It would also help us learn how these systems work, he says.

Others are calling for a new regulatory framework governing algorithms, much like what we have for the financial industry, for example. A report commissioned in 2014 by the White House recommends that policy-makers pay more attention to what the algorithms do with the data they collect and analyse. To ensure accountability, however, there would need to be independent auditors who inspect algorithms and monitor their impact. We cannot leave it to governments or industry alone to respond to the problems, says Gangadharan.

However, independent auditors face tough obstacles. For a start, digging around inside proprietary services typically violates their terms of use agreement, which prohibits attempts to analyse how they work. Under the US Computer Fraud and Abuse Act, such snooping may even be illegal. In addition, while public scrutiny is important, the details of proprietary algorithms would need to be kept safe from competitors or hackers.

The right to know

Despite these issues, in 2016 the European Parliament approved the General Data Protection Regulation (GDPR), a new set of rules governing personal data. Due to go into effect in 2018, it introduces a 'right to explanation': the opportunity for European Union citizens to question the logic of an automated decision – and contest the results.

The GDPR is a significant step forward compared with existing laws, says Bryce Goodman at the University of Oxford's Internet Institute. It creates new rules about how data is used and explicitly states how these rules affect any company working with data belonging to European citizens, whether or not that company is based in Europe. It also has power. Organizations in breach of the GDPR can expect fines of up to 4 per cent of their yearly turnover, or €20 million – whichever is greater. The GDPR also specifically calls for companies to prevent discrimination based on personal characteristics such as race, religious beliefs or health data.

The GDPR will not be easy to enforce, however. Who will explain the opaque workings of machine-learning algorithms to those who are not technically literate? But it's important to try. 'History teaches us that human decisions can all too easily be biased, whether consciously or unconsciously,' said Ed Felten, deputy chief technology officer at the White House Office of Science and Technology Policy. 'As we build automated systems, we have a responsibility to do better.'

Software watchdogs

One approach is to build software that checks other software. Kroll is working on a system that would let an auditor verify that an algorithm did what it was supposed to with what it was given. In other words, it would provide a foolproof way of checking that the outcome of the green card lottery, for example, was in fact random, or that a driverless car's algorithm for avoiding pedestrians treats both people walking and people in wheelchairs with the same caution.

Sorelle Friedler, a computer scientist at Haverford College in Pennsylvania, has a different approach. By understanding the biases inherent in the underlying data, she hopes to eliminate bias in the algorithm. Her system looks for correlations between arbitrary properties – like height or address – and demographic groupings like race or gender. If the correlation is expected to lead to unwanted bias, then it would make sense to normalize the data. It is essentially affirmative action for algorithms, she says.

That is fine for cases where discrimination is clear, when a system is found to be unfair or illegal. But what if there is disagreement about how software ought to behave? Some would argue that highly personalized price adjustment can benefit both customers and retailers. Others will defend the results given by automated services for prison sentencing. What is unacceptable to some will not be for others.

Unlike for financial systems, there are no standards of practice governing algorithms. How we want them to behave is going to be a hard question to answer. Maybe we will need AI for that.

8

Will machines inherit the Earth?

How superintelligent machines might revolutionize our world

For decades we have had machines that can out-calculate us. But the rapid pace at which AI is improving right now – and the range of tasks to which it has been successfully applied – has taken us by surprise. For some people, it is inevitable that machines will become more intelligent than us – and soon. Such superintelligent machines could revolutionize everything, from tackling climate change to health to social care. But their rise would raise tricky questions about everything, from theology to the future of our species.

The dawn of superintelligence

The notion of the superintelligent machine – one that can surpass human thinking on any subject – was introduced in 1965 by the mathematician I. J. Good, who worked with Alan Turing at Bletchley Park. Good noted that 'the first ultra-intelligent machine is the last invention that man need ever make' because, from then on, the machines would be designing other, ever-better machines.

Depending on whom you talk to, humanity's last invention could be just around the corner. The potential arrival of self-improving machines is now wrapped up in the idea of the Singularity, when artificial intelligence will outstrip our own – an event that futurists such as Ray Kurzweil believe is only a couple of decades away. Others think superintelligence – and the accompanying fears about what that would mean for humans – is a fantasy dreamed up by a culture raised on science fiction.

Sci-fi scare stories miss the point, though. We need to forget fears about AI going rogue: a world with superintelligent machines in it will be far stranger than that.

Interview: The Singularity is just around the corner

Ray Kurzweil is a computer scientist, inventor and futurist. In 2009 he co-founded the Singularity University – based at the NASA Research Park in California – specializing in 'exponential technologies'. Since 2012 he has had a full-time position at Google. He takes 150-plus supplements a day to ward off old age until the Singularity arrives.

For Kurzweil, being human with limited brainpower and a body with a use-by date is not good enough. His idea of the Singularity is that it is a near-future point in time when machine intelligence outstrips our own and AI starts to improve itself exponentially. To keep up, as he explains in this 2009 interview for New Scientist, *humans will merge with machines, become superintelligent and live for ever. From MIT to the White House, people either hate the idea or they cannot wait for it to happen.*

When will the Singularity arrive?

By 2045, give or take. We are already a hybrid of biological and non-biological technology. A handful of people have electronic devices in their brain, for example. The latest generation allows medical software to be downloaded to a computer inside your brain. But if you consider that 15 to 20 years from now these technologies will be 100,000 times smaller and a billion times more powerful, you get some idea of what will be feasible. And even though most of us don't have computers in our bodies, they are already part of who we are.

What about people who don't want to be 'transhuman' and merge with technology?

How many people completely reject all medical and health technology, don't wear glasses or take any medicine? People say they don't want to change themselves, but then when they get a disease they will do whatever they can to overcome it. We're not going to get from here to the world of 2030 or 2040 in one grand leap – we're going to get there through thousands of little steps. Put these steps together and ultimately the world is a different place.

Can we outrun our current environmental problems to reach 2045?

Yes. The resources are much greater than they appear. We only have to capture 1 part in 10,000 of the sunlight to get all the energy we need. Nanotechnology is being applied to solar energy collection technology and that is scaling up at an exponential rate. Such new technologies are ultimately very inexpensive because they are subject to the law of accelerating returns.

What do you mean by the law of accelerating returns?

The power of ideas to change the world is accelerating and few people grasp the implications of that fully. People don't think exponentially, yet exponential change applies to anything that involves measuring information content. Take genetic sequencing. When the human genome project was announced in 1990, sceptics said: 'No way you're going to do this in 15 years.' Halfway through the project the sceptics were still going strong, saying you've only finished 1 per cent of the project. But that's actually right on schedule: by the time you get to 1 per cent you're only seven doublings away.

You have a strong track record with your predictions. Has this exponential thinking helped get the timing right?

In the mid-1980s I predicted the emergence of the World Wide Web for the mid-1990s. It seemed ridiculous then, when the entire US defence budget could only link up a few thousand scientists. But I saw it doubling every year and it happened right on schedule. It is quite remarkable how predictable these measures of the power of information technology are. Even so, millions of innovators are going to come up with unexpected ideas. Who would have anticipated social networks? If 20 years ago I had said we're going to create an encyclopaedia and anybody can write and edit it, you'd have thought, my god, it's going to be full of graffiti and completely worthless. It's amazing how good it is if we harness the collective wisdom.

These advances all sound very utopian.

They are not utopian because technology is a double-edged sword; it introduces new problems as well. Overall, though, I do believe the benefits outweigh the damage that technology causes. Not everybody agrees.

Why did you set up the Singularity University?

Peter Diamandis – founder and chairman of the X Prize Foundation – and I decided the time was right to start a university to bring together the leading people in artificial intelligence, nanotechnology, biotechnology and advanced computing to help solve the problems of the future, because these problems are complex and multidimensional. NASA and Larry Page of Google are also backing it. It's a very intensive nine-week course.

You have said that you would like to bring your father back to life because you miss him.

That's right – using DNA from his grave collected by nanobots, then adding all the information extracted by AI from my memories and those of other people who remember him. Plus all the mementos of his life that I've kept, in boxes and elsewhere, could be downloaded. He could be an avatar, or a robot or in some other form.

A question of values

Concern that smart machines might do away with us has been brewing since the advent of modern computers in the 1950s, but was confined to the wilder fringes of AI. In recent years, however, a school of thought led by the philosopher Nick Bostrom has made this 'existential risk' a mainstream talking point. His 2014 book *Superintelligence* won over technocrats like Bill Gates, Elon Musk and Apple co-founder Steve Wozniak.

Public figures like Stephen Hawking joined the chorus, too. Hawking told the BBC that 'the development of full artificial intelligence could spell the end of the human race ... It would take off on its own, and redesign itself at an ever increasing rate. Humans, who are limited by slow biological evolution, couldn't compete, and would be superseded.' In 2016 he followed that up by saying that AI is likely 'either the best or worst thing ever to happen to humanity'.

One of Bostrom's more celebrated examples is of an AI bent only on making paper clips: it might use up all the planet's resources in pursuit of its objective. Alternatively, an AI tasked with making humans happy might cut out parts of our brains associated with unpleasant experiences. So the challenge is to ensure that an AI's goals are compatible with our own.

In July 2016 a few dozen researchers, philosophers and ethicists gathered at a private meeting in Jesus College, Cambridge, UK, to discuss the issues. 'Existential risk boils down to a question of values,' John Naughton, Emeritus Professor of the Public Understanding of Technology at the Open University, UK, told the meeting. For Naughton, the bad news is that those leading the AI charge typically have a technocratic attitude in which data-driven decision-making is good – end of discussion.

So how should we set goals and values for future machines? The simple answer is we do not yet know. Although current AIs are trained on data sets to perform specific tasks, their successors may be able to choose their own objectives, just as we do. They might work out better solutions to problems that way. But if we give them that freedom, we need the ability to stop them from taking undesirable paths – hence talk of 'kill switches' that could be built into future AI.

Another common value judgement is that an AI should aim for the greatest good for the greatest number. That initially sounds attractive. For example, it is more cost-effective to buy malaria nets than develop drugs for rare diseases. But that could mean abandoning the kind of individual gestures we hold dear, and which are important for social cohesion. AIs might be able to 'out-ethic us', making cold-bloodedly rational choices on our behalf, but we might not like the results. That will become more of a concern as they push even further into areas currently reserved for humans.

What happens if AI becomes smarter than we are?

Philosopher Nick Bostrom is the director of the Future of Humanity Institute at the University of Oxford and author of Superintelligence: Paths, Dangers, Strategies *(2014). He says that one day we will create AIs far superior to us. Here he explains why designing them wisely is the greatest challenge we face.*

'Humans have never encountered a more intelligent life form, but this will change if we create machines that greatly surpass our cognitive abilities. Then our fate will depend on the will of such a "superintelligence", much as the fate of gorillas today depends more on what we do than on gorillas themselves.

'We therefore have reason to be curious about what these superintelligences will want. Is there a way to engineer their motivation systems so that their preferences will coincide with ours? And supposing a superintelligence starts out human-friendly, is there some way to guarantee that it will remain benevolent even as it creates ever more capable successor-versions of itself?

'These questions – which are perhaps the most momentous that our species will ever confront – call for a new science of advanced artificial agents. Most of the work answering these questions remains to be done, yet over the last ten years, a group of mathematicians, philosophers and computer scientists have begun to make progress. As I explain in my book *Superintelligence: Paths, Dangers, Strategies*, the findings are at once disturbing and deeply fascinating. We can see, in outline, that preparation for the machine intelligence transition is the essential task of our time.

'But let us take a step back and consider why machines with high levels of general intelligence would be such a big deal. By a superintelligence I mean any intellect that greatly exceeds the cognitive performance of humans in virtually all domains. Plainly, none of our current AI programs meets this criterion. All compare unfavourably in most respects, even to a mouse.

'So we are not talking about present or near-future systems. Nobody knows how long it will take to develop machine intelligence that matches humans in general learning and reasoning ability. It seems plausible that it might take a number of decades. But once AIs do reach and then surpass this level, they may quickly soar to radically superintelligent levels.

'After AI scientists become more capable than human scientists, research in artificial intelligence would be carried out by machines operating at digital timescales, and progress would be correspondingly rapid. There is thus the potential for an intelligence explosion, in which we go from there being no computer that exceeds human intelligence to machine superintelligence that enormously outperforms all biological intelligence.

'The first AI system to undergo such an intelligence explosion could then become extremely powerful. It would be the only superintelligence in the world, capable of developing a host of other technologies very quickly, such as nanomolecular robotics, and using them shape the future of life according to its preferences.

'We can distinguish three forms of superintelligence. A speed superintelligence could do everything a human mind could do, but much faster. An intelligent system that runs 10,000 times faster than a human mind, it would be able to read a book in a few seconds and complete a PhD thesis in an afternoon. To such a fast mind, the external world would appear to run in slow motion.

'A collective superintelligence is a system composed of a large number of human-level intellects organized so that the system's performance as a whole vastly outstrips that of any current cognitive system. A human-level mind running as software on a computer could easily be copied and run on multiple computers. If each copy was valuable enough to repay the cost of hardware and electricity, a massive population boom could result. In a world with trillions of these intelligences, technological progress may be much faster than it is today, since there could be thousands of times more scientists and inventors.

'Finally, a quality superintelligence would be one that is at least as fast as a human mind and vastly qualitatively smarter. This is a more difficult notion to comprehend.

The idea is that there might be intellects that are cleverer than humans in the same sense that we are cleverer than other animals. In terms of raw computational power, a human brain may not be superior to, say, the brain of a sperm whale, possessor of the largest known brain, weighing in at 7.8 kg compared to 1.5 kg for an average human. And, of course, the non-human animal's brain is nicely suited to its ecological needs.

'Yet the human brain has a facility for abstract thinking, complex linguistic representations and long-range planning that enables us to do science, technology and engineering more successfully than other species. But there is no reason to suppose that ours are the smartest possible brains. Rather, we may be the stupidest possible biological species capable of starting a technological civilization. We filled that niche because we got there first – not because we are in any sense optimally adapted to it.

'These different types of superintelligence may have different strengths and weaknesses. For example, a collective superintelligence would excel at problems that can be readily subdivided into independent subproblems, whereas a quality superintelligence may have an advantage on problems that require new conceptual insights or complexly coordinated deliberation.

'The indirect reaches of these different kinds of superintelligence, however, are identical. Provided the first iteration is competent in scientific research, it is likely to quickly become a fully general superintelligence. That's because it would be able to complete the computer or cognitive science research and software engineering needed to build for itself any cognitive faculty it lacked at the outset.

'Once developed to this level, machine brains would have many fundamental advantages over biological brains, just as engines have advantages over biological muscles. When it comes to the hardware, these include vastly greater numbers of processing elements, faster frequency of operation of those elements, much faster internal communication and superior storage capacity.

'Advantages in software are harder to quantify, but they may be equally important. Consider, for example, copyability. It is easy to make an exact copy of a piece of software, whereas "copying" a human is a slow process that fails to carry over to the offspring the skills and knowledge that its parents acquired during their lifetimes. It is also much easier to edit the code of a digital mind: this makes it possible to experiment and to develop improved mental architectures and algorithms. We are able to edit the details of the synaptic connections in our brains – this is what we call learning – but we cannot alter the general principles on which our neural networks operate.

'We cannot hope to compete with such machine brains. We can only hope to design them so that their goals coincide with ours. Figuring out how to do that is a formidable problem. It is not clear whether we will succeed in solving that problem before somebody succeeds in building a superintelligence. But the fate of humanity may depend on solving these two problems in the correct order.'

Super-creators

A glimpse at the new kinds of challenge we can expect AI to start tackling came just a few months before the 2016 Cambridge meeting. AlphaGo's win against world champion Lee Sedol at the hugely complex board game Go was reminiscent of Garry Kasparov's 1997 bouts with IBM's Deep Blue supercomputer. But whereas that joust demonstrated machines' superiority at brute-force calculation, AlphaGo's victory showed something else: creativity and intuition. At the meeting, Demis Hassabis, co-founder of AlphaGo's creator DeepMind, proposed that creativity be defined as the ability to synthesize knowledge to produce a novel idea, and that intuition is implicit knowledge acquired through experience that is not consciously expressible.

AlphaGo won one match by playing a move that departed from centuries of received wisdom. It cannot express why it did this, but clearly had a rationale. So was it being creative and intuitive, albeit in a very limited way? If so, it might represent a new class of smart machine: 'super-creators', say, rather than supercomputers.

But it is missing the point to describe creativity as an innate property, said Simon Colton, who studies computational creativity at Goldsmiths, University of London. While he looks forward to a future in which, say, your phone endlessly composes music, he says creativity is a social construct conferred on a person – or entity – by others. Colton has made machines that paint pictures and make up storylines, but says it can be impossible to evaluate computer-generated works without imposing an invalid human frame on them.

What about qualities we still think of as exclusively human – imagination, emotion and, above all, consciousness? Machines that probe these areas are in development, but the AIs hitting the headlines do not get anywhere near them. While a system can be trained to perform new tasks, it cannot usually transfer knowledge gained in one area to another, as humans do.

The spectrum of future machines

Many researchers agree that the way most people imagine AI – a machine that thinks just like a human – is a remote prospect, unlikely to be fulfilled without a better understanding of how our own minds work. There is a broad consensus that such artificial general intelligence (AGI) is achievable this century, but few believe it will result from just carrying on as we have so far. The field has a history of AI winters, when development grinds to a halt after a period of rapid advance.

A superintelligent machine need not replicate all facets of humanity (see Figure 8.1). The spectrum of future machines could include 'zombie' AGIs that resemble humans but have no consciousness, says Murray Shanahan, who studies cognitive robotics at Imperial College London – as well as AIs that are more conscious than us. That puts them into company with extraterrestrial intelligences, which might also be super-smart but utterly inhuman.

There is a final conundrum: how would the creation of machines as intelligent and/or conscious as ourselves challenge our ideas about our place in the cosmos? Perhaps surprisingly, the religious might need less adjustment. Abrahamic faiths, at least, need not have a problem with there being non-human intelligences, said Cambridge theologian Andrew Davison, given that the Golem of Jewish folklore and immaterial super-beings in the Bible are examples of just that.

We have only just begun to live with smart machines. While we worry today about killer robots, the challenges to come may be turn out to be much stranger. One day, we may find ourselves living alongside aliens and angels.

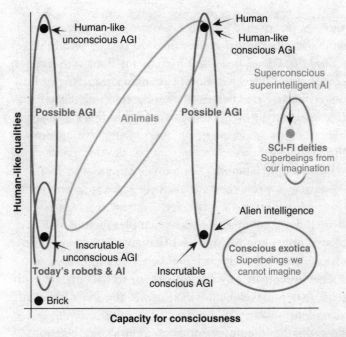

FIGURE 8.1 Super-smart machines – known as artificial general intelligence (AGI) – do not have to think like us or share human-like traits such as empathy.

Can software suffer?

Computational neuroscientist and futurist Anders Sandberg researches the ethical and social issues around human enhancement and new technologies at the Future of Humanity Institute, University of Oxford. One day we will create virtual minds, he says. Could they feel pain?

'I turned off the computer as I left my office, deleting the neural network simulation that I had been working on. Then a thought hit me: had I just killed something? I rationalized that the simulation was simpler than the systems within the bacteria I was doubtlessly squashing on the floor. If they did not matter, neither did it. But the doubt remained ...

'Science has a problem. If we want to find out what really happens in living beings, or how to cure diseases, often we must experiment on them. Digital simulations offer a way out.

'Ever since the squid giant axon was modelled in the 1950s using a mechanical calculator, our ability to simulate biological systems has skyrocketed. Today we can run neural simulations on supercomputers that contain hundreds of millions of realistic neurons and billions of synapses. Cells and their chemistry have been modelled to a similar extent.

'This potentially offers an alternative to animal experiments. Instead of subjecting a living creature to pain when testing a painkiller, why not simulate the pain system and check whether the treatment works? The logical endpoint of this is an emulation, where every part of the brain – and body – is simulated digitally.

'The challenge is to map the connectivity in real brains. It will be years before we can create even a proper insect brain, but work is already under way to create the brain of the nematode worm *C. elegans* in virtual form. The worm is a good candidate for this because it has one of the simplest brains of any organism, with just 302 nerve cells. In 2012 researchers at the University of Waterloo, Canada, announced the creation of a large functional brain simulation, SPAUN, with 2.5 million neurons. And the Human Brain Project, a European collaboration, has the ultimate aim of simulating a whole human brain.

'Although these digital emulations could resolve many existing ethical dilemmas, they raise new ones. The first is that many real animals must be sacrificed to create a virtual one. We may one day scan the final lab rat, which will become Standard Lab Rat 1.0, and rely on simulation from then on – but there will have been years of basic neuroscience to enable that simulation. The second problem is that we need to be certain our simulations are right if we want to trust them with our drug testing or other research.

'It is the third problem that really interests me. Would emulations feel pain? Do we have to care for them like we do for animals or humans involved in medical research? This hinges on whether software can suffer. For example, Sniffy the Virtual Rat lets users observe the behaviour of rats given electric shocks to teach the psychology of learning without using live animals. Yet few of us would think there is any real pain there: it is essentially an interactive cartoon, similar to a virtual pet toy. We may empathize with it, but it is similar to a talking doll. Whole brain emulations, which recreate the neural connections of animals and even humans, are a different matter.

'In his 1978 paper "Why you can't make a computer that feels pain", philosopher Daniel Dennett argued that we don't have a rigorous enough definition of pain, so we cannot build a machine that feels it. But he also believed that we may eventually figure it out, and at some point thoughtful people would refrain from kicking robots. Other philosophers, such as John Searle, have argued that, no matter how sophisticated the simulation is, it will always be mere numbers updated in complicated ways: there cannot be real intentions or consciousness in pure software. It might also be that the system needs to have a body to ground it in the real world.

'But what about CyberChild, created by the neuroscientist Rodney Cotterill as a model of his theory of consciousness? It is a virtual simulated infant with a brain and body model based on real biology. It has internal states such as blood sugar levels, and activity in different brain areas. It responds to these internal states, it can learn, it needs food – if its nutrient levels are too low it "dies" – and it can cry and flail its arms. Yes, it is a very simple organism, but it is intended to be conscious. There is something eerie about it: assuming Cotterill's theory is right, in principle this being could have experience.

'We know brains exist for motivating actions that lead to better outcomes for the organism: this is the whole point of pain, pleasure and planning. If we were to make a perfect copy of the activity of a brain, we would get the same behaviour, based on the same pattern of internal interactions. There is no way of telling from the outside whether it has any real experience, whatever that is. There is considerable disagreement about whether software can suffer, or whether it matters morally. So what should we do?

'My suggestion is that it is better to be safe than sorry: assume that any emulated system could have the same mental properties as the organism or biological system it is based on, and treat it accordingly. If your simulation just produces neural noise, you have a good reason to assume there is nothing in there to care about. But if you make an emulated mouse that behaves like a real one, you should treat it as you would treat a lab mouse.

'I agree this is inconvenient for computational neuroscience. But it is probably the moral thing to do. Once we get to simulated vertebrates, we ought to apply government animal-testing guidelines. We should avoid generating virtual suffering by not running experiments that produce pain signals.

'But we can also improve on biology, because in simulations we can leave out pain systems, simulate perfect side-effect-free painkillers or just block neural activity related to suffering. We could in principle monitor the emulated brain for any kind of suffering and stop as soon as we detected it. There is also the issue of quality of life. We have begun to recognize that giving animals good environments matters – building equally good virtual environments may prove a hurdle. Virtual rats would plausibly need virtual fur, whiskers and smells to feel at home.

'What about euthanasia? Living organisms die permanently, and death means the loss of their only chance at being alive. But an emulated brain could be restored from a backup: Lab Rat 1.0 would awake in the same way no matter how many copies had been tested in the past. The only thing lost when restoring it would be the memories of the previous experiment. There may still be pleasures and pains that count. In some ethical views, running a million supremely happy rat simulations in the background might be a "moral offset" for doing something painful to one.

'In the long run I believe we will create human brain emulations. Their moral status will in many ways be easier to determine than for animals: just ask them. Take an eminent philosopher doubting software can be conscious, scan their brain, and ask the resulting emulation if it feels conscious. If the response is: "… yes. Darn. I need to write a paper!" we have pretty good evidence that there is a being with enough intellect, introspection and moral value to deserve rights. But until then we should treat our software animals well. Just in case.'

Computers that defy logic

For 75 years computers have worked within limits defined by Alan Turing. These limits may place limits on how smart artificial intelligence can get. But work has begun to fulfil Turing's prophecy of a machine that can solve the unsolvable, which he called the 'oracle'. In his PhD thesis of 1938, Turing specified no further what shape it might take. Perhaps that is fair enough: aged just 26, he had already lit the fuse of a revolution. So absorbed have we been in exploring his rich and varied legacy, and transforming our world with the machines and applications that built upon it, that we have rather overlooked the oracle. Turing had shown with his universal machine that any regular computer would have inescapable limitations. With the oracle, he showed how you might smash through them.

In his short life, Turing never tried to turn the oracle into reality. Perhaps with good reason: most computer scientists believe anything approximating an oracle machine would soon fall foul of fundamental restrictions on how information and energy flow in the universe. You could never actually make one.

In a lab in Springfield, Missouri, two researchers are seeking to prove the sceptics wrong. Building on theoretical and experimental advances of the past two decades, Emmett Redd and Steven Younger of Missouri State University think a 'super-Turing' computer is within our grasp. With it, they hope, could come insights not just into the limits of computation in the cosmos but into the most intriguingly powerful computer we know of within it: the human brain.

Computers as we know them are in essence very capable, rigorous and efficient renderings of what we humans might be capable of given precise instructions, a high boredom threshold and a limitless supply of paper and pencils. They excel at successive additions, multiplications, logical decisions; if x then y. A universal computing machine – often known simply as a Turing machine – does the same things, only without the tedium. 'Electronic computers are intended to carry out any definite rule-of-thumb process which could have been done by a human operator working in a disciplined but unintelligent manner,' as Turing himself wrote in the programmer's handbook for the University of Manchester's Mark II computer in 1950.

Computers, then, have their blind spots just as we do. No matter how disciplined, well schooled or patient we are, certain questions defy our logic. What is the truth of the statement 'This statement is false'? In 1931 the mathematician Kurt Gödel demonstrated that this problem was universal, with his infamous incompleteness theorems, showing that any system of logical axioms would always contain such unprovable statements.

Similarly, as Turing showed, a universal computer built on logic alone always encounters 'undecidable' problems that never yield straight answers, no matter how much processor power you throw at them. An oracle as Turing envisaged it was essentially a black box whose unspecified contents would be able to solve undecidable problems. An 'O-machine', he proposed, would exploit whatever was in this black box to go beyond the bounds of conventional human logic – and so surpass the abilities of every computer ever built.

That is as far as he went in 1938. Then, more than 50 years later, Hava Siegelmann came up with a model for a super–Turing computer by accident. In the early 1990s she was working on her PhD in neural networks at Rutgers University in Piscataway, New Jersey, just a 40-minute drive from Princeton, where Turing had presented his thesis. Siegelmann's initial aim was to prove theoretically the limits of neural networks: to show that, for all their flexibility, they could never have the full logical capabilities of a conventional Turing machine. She failed time and again. Eventually, she proved the reverse. One of the hallmarks of a Turing machine is that it is incapable of generating true randomness. By weighting a network with the infinite, non-repeating number strings of irrational numbers such as pi, Siegelmann showed that you could, in theory, make it super–Turing. In 1993 she even showed how such a network could solve the halting problem.

Her fellow computer scientists met the idea with coolness, and in some cases downright hostility. Various ideas had been floated for 'hypercomputers' that might exploit exotic physics to go super–Turing, but they always seemed to lie on a scale from implausible to ridiculous. Siegelmann eventually published her proof in 1995, but she soon lost interest, too. 'I believed it was mathematics only, and I wanted to do something practical,' she says. 'I turned down giving any more talks on super–Turing computation.'

Redd and Younger had been aware of Siegelmann's work for a decade before they realized that their own research was heading in the same direction. In 2010 they were building neural networks using analogue inputs that, unlike the conventional digital code of 0 and 1, can take a potentially infinite range of values between the two. There was more than a trace of Siegelmann's endless irrational numbers in there.

Powered by chaos

In 2011 they approached Siegelmann, by then director of the Biologically Inspired Neural & Dynamical Systems Laboratory at the University of Massachusetts in Amherst, to see whether she might be interested in collaboration. She said yes. As it happened, she had recently started thinking about the problem again, and was beginning to see how irrational-number weightings were not the only project worth addressing. Anything that introduced a similar element of randomness or unpredictability might do the trick, too.

The route the trio chose was chaos. A chaotic system is one whose response is very sensitive to small changes in its initial conditions. Wire up an analogue neural net in the right way, and tiny gradations in its outputs can be used to create bigger changes at the inputs, which in turn feed back to cause bigger or smaller changes, and so on. In effect, the system becomes driven by an unpredictable, infinitely variable noise.

The researchers are working on two small prototype chaotic machines. One is a neural network based on standard electronic components, with three 'neurons' in the form of integrated circuit chips and 11 synaptic connections on a circuit board a little larger than a hardback book. The other, with 11 neurons and around 3,600 synapses, uses lasers, mirrors, lenses and photon detectors to encode its information in light.

Although only on a small scale, that should be enough, the team thinks, to take them beyond Turing computation. It is a claim that invites plenty of scepticism. The main concern is that mathematical models involving any sort of infinity always run into problems when they are forced to deal with reality. It is not that the maths does not work – it is just a moot point whether true randomness is something we can harness, or whether it even exists.

That question was clearly on Turing's mind: he often speculated about a connection between intrinsic randomness and the origin of creative intelligence. In 1947 he went so far as to suggest to his astounded bosses at the UK National Physical Laboratory near London that they should put radioactive radium into the Automatic Computing Engine he had devised, in the hope that its seemingly random decays would give its inputs the desired unpredictability. 'I don't think he intended to build the oracle machine,' says Siegelmann. 'What he had in mind was to build something that's more like the brain.'

Since then, building a computer with brain-like qualities has been a perennial aim, with the latest large-scale initiative being part of the Human Brain Project based at the Swiss Federal Polytechnic School in Lausanne. These endeavours, though, are all about building replica neurons with standard, digital Turing-machine technology. Younger and Redd are aware of the difficulties but are still convinced that the less rigid approach of their chaotic neural networks is more likely to bear fruit.

Interview: Why AI is a dangerous dream

Noel Sharkey is Emeritus Professor of Artificial Intelligence and Robotics at the University of Sheffield, a co-founder of the International Committee for Robot Arms Control and a leading member of the Campaign to Stop Killer Robots. In this 2009 interview for New Scientist, *he explains why he worries that AI is a dangerous myth that could lead to a dystopian future of unintelligent, unfeeling robot carers and soldiers.*

What does artificial intelligence mean to you?

I like AI pioneer Marvin Minsky's definition of AI as the science of making machines do things that would require intelligence if done by humans. However, some very smart human things can be done in dumb ways by machines. Humans have a very limited memory, so, for us, chess is a difficult pattern-recognition problem that requires intelligence. A computer like Deep Blue wins by brute force, searching quickly through the outcomes of millions of moves. It is like arm-wrestling with a mechanical digger. I would rework Minsky's definition as the science of making machines do things that lead us to believe they are intelligent.

Are machines capable of intelligence?

If we are talking intelligence in the animal sense, I would have to say no. For me, AI is a field of outstanding engineering achievements that helps us to model living systems but not replace them. It is the person who designs the algorithms and programs the machine who is intelligent, not the machine itself.

Are we close to building a machine that can meaningfully be described as sentient?

I'm an empirical kind of guy, and there is just no evidence of an artificial toehold in sentience. It is often forgotten that the idea of mind or brain as computational is merely an assumption, not a truth. When I point this out to 'believers' in the computational theory of mind, some of their arguments are almost religious. They say, 'What else could there be? Do you think mind is supernatural?' But accepting mind as a physical entity does not tell us what kind of physical entity it is. It could be a physical system that cannot be recreated by a computer.

So why are predictions about robots taking over the world so common?

There has always been fear of new technologies based upon people's difficulties in understanding rapid developments. I love science fiction and find it inspirational, but I treat it as fiction. Technological artefacts do not have a will or a desire, so why would they 'want' to take over? Isaac Asimov said that, when he started writing about robots, the idea that robots were going to take over the world was the only story in town. Nobody wants to hear otherwise. I used to find, when newspaper reporters called me and I said I didn't believe AI or robots would take over the world, they would say thank you very much, hang up and never report my comments.

You describe AI as the science of illusion.

It is my contention that AI, and particularly robotics, exploits natural human zoomorphism. We want robots to appear like humans or animals, and this is assisted by cultural myths about AI and a willing suspension of disbelief. The old automata makers, going back as far as Hero of Alexandria, who made the first programmable robot in 60 CE, saw their work as part of natural magic – the use of trick and illusion to make us believe their machines were alive. Modern robotics preserves this tradition with machines that can recognize emotion and manipulate silicone faces to show empathy. Chatbots are adept at finding conversationally appropriate sentences. If AI workers would accept the trickster role and be honest about it, we might progress a lot more quickly.

These views are in stark contrast to those of many of your peers in the robotics field.

Yes. Roboticist Hans Moravec says that computer-processing speed will eventually overtake that of the human brain and make them our superiors. The inventor Ray Kurzweil says humans will merge with machines and live forever by 2045. To me these are just fairy tales. I don't see any sign of it happening. These ideas are based on the assumption that intelligence is computational. It might be, and equally it might not be. My work is on immediate problems in AI, and there is no evidence that machines will ever overtake us or gain sentience.

Do you believe that there are dangers if we fool ourselves into believing the AI myth?

It is likely to accelerate our progress towards a dystopian world in which wars, policing and care of the vulnerable are carried out by technological artefacts that have no possibility of empathy, compassion or understanding.

How would you feel about a robot carer looking after you in old age?

Eldercare robotics is being developed quite rapidly in Japan. Robots could be greatly beneficial in keeping us out of care homes in our old age, performing many dull duties for us and aiding in tasks that failing memories make difficult. But it is a trade-off. My big concern is that, once the robots have been tried and tested, it may be tempting to leave us entirely in their care. Like all humans, the elderly need love and human contact, and this often only comes from visiting carers. A robot companion would not fulfil that need for me.

You also have concerns about military robots.

The many thousands of robots in the air and on the ground are producing great military advantages. No one can deny the benefit of their use in bomb disposal and surveillance to protect soldiers' lives. My concerns are with the use of armed robots. Drone attacks are often reliant on unreliable intelligence in the same way as in Vietnam, where the US ended up targeting people who were owed gambling debts by its informants. This over-reach of the technology is killing many innocent people. US planning documents make it clear that there is a drive towards developing autonomous killing machines. There is no way for an AI to discriminate between a combatant and a civilian. Claims that such a system is coming soon are unsupportable and irresponsible.

Is this why you are calling for ethical guidelines and laws to govern the use of robots?

In the areas of robot ethics that I have written about – childcare, policing, military, eldercare and medical – I have spent a lot of time looking at current legislation around the world and found it wanting. I think there is a need for urgent discussions among the various professional bodies, the citizens and the policy-makers to decide while there is still time. These developments could be upon us as fast as the Internet was, and we are not prepared. My fear is that once the technological genie is out of the bottle it will be too late to put it back.

Five reasons why the Singularity will never arrive

Stephen Hawking – who has little expertise in AI – has report-edly toned down his talk of an AI apocalypse after debating the issue with DeepMind's Demis Hassabis. But the fears expressed by the likes of Hawking and Bill Gates revolve around the idea of the Singularity. At that point – the argument goes – a more intelligent species starts to inhabit the Earth. We can trace the idea back to a number of different thinkers, including John von Neumann, one of the founders of computing, and the science-fiction author Vernor Vinge.

The idea has been around almost as long as AI itself. In 1958 mathematician Stanisław Ulam wrote a tribute to the recently deceased von Neumann, in which he recalled: 'One conver-sation centred on the ever accelerating progress of technol-ogy and changes in the mode of human life, which gives the appearance of approaching some essential singularity … beyond which human affairs, as we know them, could not continue.'

There are several reasons to be fearful of machines overtak-ing us in intelligence. Humans have become the dominant spe-cies on the planet largely because we are so intelligent. Many animals are bigger, faster or stronger than us. But we used our intelligence to invent tools, agriculture and amazing technolo-gies like steam engines, electric motors and smartphones. These have transformed our lives and allowed us to dominate the planet. It is therefore not surprising that machines that think – and might even think better than us – threaten to usurp us. Just as elephants, dolphins and pandas depend on our goodwill for their continued existence, our fate in turn may depend on the decisions of these superior thinking machines.

The idea of an intelligence explosion, when machines recursively improve their intelligence and thus quickly exceed human intelligence, is not a particularly wild idea. The field of computing has profited considerably from many similar exponential trends. So it is not unreasonable to suppose AI will also experience exponential growth. But there are several strong reasons why the Singularity is improbable.

1 The 'fast-thinking dog' argument

Silicon has a significant speed advantage over our brain's wetware and this advantage doubles every two years or so, according to Moore's law. But speed alone does not bring increased intelligence. Even if I can make my dog think faster, it is still unlikely to play chess. It does not have the necessary mental constructs, the language and the abstractions. Steven Pinker put this argument eloquently: 'Sheer processing power is not a pixie dust that magically solves all your problems.'

Intelligence is much more than thinking faster or longer about a problem than someone else. Of course, Moore's law has helped AI. We now learn faster, and off bigger data sets. Speedier computers will certainly help us to build AI. But, at least for humans, intelligence depends on many other things including years of experience and training. It is not at all clear that we can short-circuit this in silicon simply by increasing the clock speed or adding more memory.

2 The anthropocentric argument

The Singularity supposes that human intelligence is some special point to pass, some sort of tipping point. If there is one thing that we should have learned from the history of science, it is that we are not as special as we would like to believe. Copernicus taught us that the universe does not revolve around the Earth. Darwin showed us that we are not so different from other apes. Watson, Crick and Franklin revealed that the same DNA code of life powers us and the simplest amoeba. And artificial intelligence will no doubt teach us that human intelligence is itself nothing special. There is no reason to suppose that human intelligence is a tipping point that, once passed, allows for rapid increases in intelligence.

Of course, human intelligence is a special point because we are, as far as we know, unique in being able to build artefacts that amplify our intellectual abilities. We are the only creatures on the planet with sufficient intelligence to design new intelligence, and this new intelligence will not be limited by the slow process of human reproduction and evolution. But that does not bring us to the tipping point, the point of recursive self-improvement. We have no reason to suppose that human intelligence is enough to design an artificial intelligence that is sufficiently intelligent to be the starting point for a technological singularity.

Even if we have enough intelligence to design super-human artificial intelligence, the result may not be adequate to precipitate the Singularity. Improving intelligence is far harder than just being intelligent.

3 The 'diminishing returns' argument

The idea of the Singularity assumes that improvements to intelligence will be by a relative constant multiplier, each generation getting some fraction better than the last. However, the performance of most of our AI systems has so far been that of diminishing returns. Much may be easily achieved at the start, but we then run into difficulties when looking for improvements. This helps explain the overly optimistic claims made by many of the early AI researchers.

An AI system may be able to improve itself an infinite number of times, but the extent to which its intelligence changes overall could be limited. For instance, if each generation only improves by half the last change, then the system will never get beyond doubling its overall intelligence.

4 The 'limits of intelligence' argument

There are many fundamental limits within the universe. Some are physical: you cannot accelerate past the speed of light, know both position and momentum with complete accuracy, or know when a radioactive atom will decay. Any thinking machine that we build will be limited by these physical laws. Of course, if that machine is electronic or even quantum in nature, these limits are likely to be beyond the biological and chemical limits of our human brains.

Nevertheless, AI may well run into some fundamental limits. Some of these may be due to the inherent uncertainty of nature. No matter how hard we think about a problem, there may be limits to the quality of our decision-making. Even a super-human intelligence is not going to be any better than you at predicting the result of the next EuroMillions lottery.

5 The 'computational complexity' argument

Finally, computer science already has a well-developed theory of how difficult it is to solve different problems. There are many computational problems for which even exponential improvements are not enough to help us solve them practically. A computer cannot analyse some code and know for sure whether it will ever stop – this is known as the 'halting problem'.

Alan Turing famously proved that such a problem is not computable in general, no matter how fast or smart we make the computer analysing the code. Switching to other types of device like quantum computers will help. But these will only offer exponential improvements over classical computers, which is not enough to solve problems like Turing's halting problem. And the hypothetical hyper-computers that might break through such computational barriers remain controversial.

Is winter coming?

Here is a prediction that might sound familiar:

> In from three to eight years, we will have a machine with
> the general intelligence of an average human being.
> I mean a machine that will be able to read Shakespeare,
> grease a car, play office politics, tell a joke, have a fight. At
> that point the machine will begin to educate itself with
> fantastic speed. In a few months it will be at genius level,
> and a few months after that, its powers will be incalculable.

It does not come from an AI visionary *du jour* such as Nick Bostrom or Elon Musk, however. It was made in 1970 by one of the fathers of AI – Marvin Minsky. But, eight years later, the cutting edge was still only the Speak & Spell, an educational toy that used rudimentary computer logic. When the chasm between Minsky's promise and reality sank in, the disappointment destroyed AI research for decades.

Today there are whispers that something similar might be on its way, fuelled by the excitement surrounding deep learning. 'I can feel the cold breeze on the back of my neck,' says Roger Schank at Northwestern University in Evanston, Illinois. But are these the grumblings of veterans who missed out on the true AI revolution, or harbingers of something real?

The original AI winter was brought about by two factors. First, a research monoculture focused on a technique called rule-based or symbolic learning, which tried to emulate basic human reasoning. This showed great promise in the lab, hence the breathless predictions of Minsky and others. As these prognostications piled up, the UK Science Research Council commissioned a report to evaluate the claims. The result was damning. The Lighthill Report of 1973 revealed that, for all the potential that rule-based learning showed in lab problems, these were all it could handle. In reality, it was undone by complexity. Governments stopped funding university AI research. Graduate students sought greener pastures in disciplines that garnered more respect. The remaining scientists talked about their work in hushed tones and deliberately eschewed the phrase 'artificial intelligence'. It would be another two decades before the field recovered.

The rehabilitation began in 1997 when IBM's Deep Blue AI defeated the reigning chess champion. In 2005 an autonomous car drove itself for 131 miles. In 2011 IBM's Watson defeated two human opponents on the game show *Jeopardy!* But what catapulted AI into the mainstream was deep learning.

An AI gold rush

In 2012, to much fanfare, Google unveiled a neural network that was able to recognize cat faces in videos (see Figure 8.2). People began to talk about how deep learning, given enough processing power, would lead to a machine able to develop concepts, and thus an understanding of the world. Two years later Google bought DeepMind, the firm that went on to win at Go, for $500 million.

These early successes have sparked an AI gold rush – based on some bold claims. One start-up promises to turn cancer into a manageable long-term disease rather than an outright killer, another wants to reverse ageing, while another has ambitions to predict future terrorists by their facial features. What unites

FIGURE 8.2 A neural network trained to spot cats in YouTube videos
marked the start of the current AI gold rush.

them is the idea that, given the right combination of algorithms,
some solution to these so far intractable problems will pop out.

'The black magic seduction of neural networks has always been
that, by some occult way, they will learn from data so they can
understand things they have never seen before,' says Mark Bishop
at Goldsmiths, University of London. Their complexity helps people suspend disbelief and imagine that the algorithms will converge to form some kind of emergent intelligence. But it is still just
a machine built on rule-based mathematical systems, says Schank.

In 2014 a paper that could be seen as the successor to the
Lighthill Report punctured holes in the belief that neural networks do anything even remotely akin to actual understanding.
Instead, they recognize patterns, finding relationships in data
sets that are so complex that no human can see them. This matters because it disproves the idea that they could develop an
understanding of the world. A neural network can say a cat is a
cat, but it has no concept of what a cat is.

The paper is not the only thing giving people déjà vu. Schank and others point to the money pouring into deep learning and the funnelling of academic talent. 'When the field focuses too heavily on short-term progress by only exploring the strength of a single technique, this can lead to a long-term dead end,' says Kenneth Friedman, a student at the Massachusetts Institute of Technology, who adds that the AI and computer science students around him are flocking to deep learning.

It's not just the old guard that is worried. Dyspeptic rumblings are coming from the vanguard of machine-learning applications, including Crowdflower, a data-cleaning company, which wonders whether AI is suffering from 'hyper-hype'.

But this fear that the AI bubble is about to pop – again – is not the mainstream view. 'I don't think it's clear that there is a bubble,' says Miles Brundage at the Future of Humanity Institute's new Strategic AI Research Centre in Oxford. Even if there is, he thinks the field is still safe for the moment. 'I don't think we're likely to see it run out of steam any time soon. There's so much low-hanging fruit, and excitement and new talent in the field,' he says.

Even the Cassandras insist they don't want to undersell it. 'I'm impressed by what people have achieved,' says Mark Bishop. 'I never thought I'd ever see them crack Go. And face recognition is at nearly 100 per cent.' But these applications are not what has everyone so excited. Instead, it is the lure of curing cancer and ending ageing. Even if AI can meet these expectations, there are still obstacles. People tend not to acknowledge how inefficient deep learning is, for example. Neither do they admit how difficult it is to get enough data to meet the claims some firms are making, especially in medicine, where privacy concerns prove a huge obstacle to obtaining sufficient amounts of big data.

It is hard to say whether this shortfall will herald a proper winter; there have been periods of disillusionment in the past without winter setting in. It is likely to depend on how much disappointment people, and funding bodies, are able to tolerate. To most in the field, though, this seems not the time to be worrying about an AI winter. In fact, AI's main problem currently seems to be that investors can't print money fast enough for the gold rush. But don't say you haven't been warned.

Cool things we do with computers

Why do we believe machines are on the verge of understanding the world around them? It may come down to the metaphors we use: machine learning, deep learning, neural networks and cognitive computing, which all suggest thinking.

'Cognition means thinking. Your machine is not thinking,' says Roger Schank at Northwestern University in Illinois. 'When people say AI, they don't mean AI. What they mean is a lot of brute force computation.' Patrick Winston at the Massachusetts Institute of Technology describes such terms as 'suitcase words': definitions so general that any meaning can be packed into them. Artificial intelligence is the prime example. Machine learning is similar – it doesn't mean learning in the traditional sense. And while there are some parallels between the two, neural networks are not neurons.

It is not just semantics. Tell people a machine is thinking and they will assume it is thinking in the way they do. If the mismatch happens enough, it could pop the AI bubble. 'The beginning and the end of the problem is the term "AI",' says Schank. 'Can we just call it "cool things we do with computers"?'

Conclusion

In 1997 IBM's Deep Blue beat Garry Kasparov at chess – then seen as the gold standard of human intellect. At the time, the defeat came as a blow. If machines were better than us at chess, what would come next? Very little, it turned out. A program that could outmatch us at chess by crunching through millions of possible moves using sheer computational calculations was not much good at anything else.

In 2016 a new breed of AI with sobering versatility and the ability to learn on the job beat Lee Sedol at Go. And this time the implications were very different. Machine-learning software seems poised to take over a wide range of human tasks, potentially putting huge numbers of people out of work and forcing us to face up to thorny ethical questions about how we want the world to operate. An unlikely alliance of philosophers, technologists and movie makers has also stoked fears that the next generation of AI might snuff out humanity.

The paranoia is perhaps not surprising. AI presents a challenge to long-standing ideas about human exceptionalism, which survived the Copernican and Darwinian revolutions but may be fatally undermined by intelligent machines. A form of techno-pessimism may also be at play: we can foresee the potential downsides but the upsides are not yet clear.

A reality check is needed. We are nowhere near the creation of a machine that replicates the full suite of a human's intellectual capabilities. And the threat of extinction by super-intelligences, if and when they arrive, is only one of a number of esoteric possibilities.

And yet, as AI becomes an ever more powerful tool, we certainly face new responsibilities. Even without a Singularity, AI could further increase the inequalities we see in society today and destabilize the current world order. One issue concerns the haves and have-nots. The best AIs in the world today are in the hands of private companies. Google has said that any truly dramatic advance will be shared with the United Nations – but under what conditions?

Many predictions about the impact AI will have on society also assume that advances will continue to arrive as quickly as they have come over the last few years – or more quickly. But that is not a given. It may be that changes happen more slowly than people expect. That is not a reason to avoid planning ahead, however. If we get it right, AI will make us all healthier, wealthier and wiser. If we get it wrong, it could be one of the worst mistakes we ever get to make.

Fifty ideas

This section helps you to explore the subject in greater depth, with more than just the usual reading list.

Four quotes about AI

1 'Machines take me by surprise with great frequency.' (Alan Turing, 1950)

2 'The question of whether machines can think is about as relevant as the question of whether submarines can swim.' (Edsger Dijkstra (1930–2002), a computer scientist who pioneered many areas of the field, 1984)

3 'Does God exist? I would say, "Not yet."' (Ray Kurzweil, an inventor and futurist, 2011)

4 'I don't work on preventing AI from turning evil for the same reason that I don't work on combating overpopulation on the planet Mars.' (Andrew Ng, a computer scientist at Stanford University and former chief scientist at China's Internet giant Baidu, 2015)

Ten Twitterbots to follow

Some estimates suggest that as many as a quarter of all tweets are generated by bots. Here are ten actually worth following.

1 **@oliviataters** is an imitation teenage girl that engages with her followers.

2 **@TwoHeadlines** tweets mash-ups of different news headlines.

3 **@haikud2** identifies tweets that fit into a haiku format.

4 **@earthquakebot** tracks earthquakes happening around the world.

5 **@valleyedits** sends an alert when someone inside Google, Facebook, Apple, Twitter or the Wikimedia Foundation makes an anonymous Wikipedia edit.

6 **@parliamentedits** does the same for anonymous edits made by someone inside the UK Parliament (there are similar bots doing this for other countries, including the US, Canada and Sweden).

7 **@greatartbot** produces original pixel art four times a day.

8 **@ArtyOriginals** retweets original artworks by the bots @ArtyAbstract, @ArtyPetals, @ArtyFractals, @ArtyMash, @ArtyShapes and @ArtyWinds.

9 **@archillect** tweets images it discovers online that it 'likes' (@archillinks follows up by tweeting picture credits for the images).

10 **@NS_headlines** generates fake article ideas for *New Scientist*.

Four AI creations to enjoy

1 **A card game invented by GenoCard** researchers at the IT University of Copenhagen in Denmark have created an AI that generates rules for new card games. Here are the rules for a three-player game called Pay the Price.

   i   The game begins with the dealer giving nine cards and 99 tokens to each player. The remainder of the deck is placed in the middle of the table.

   ii Each player then makes a mandatory bet of one or more tokens.

   iii Each player then takes one card from the deck and shows it to the other players.

   iv Each player can then take further cards from the deck, if they want, without showing the other players. But for every card taken, the player must discard three cards from their current hand.

   v   Players can repeat the preceding until they have fewer than three cards left.

   vi Once all players are happy with their hand, they reveal their cards. Ace, Jack, King, and Queen are valued as 10. The player with the highest combination wins the round and takes all tokens on the table.

'The player might notice a certain similarity to blackjack,' say GenoCard researchers José María Font and Tobias Mahlmann. The rules of blackjack were part of the initial gene pool that seeded the evolutionary algorithm that produced the game. 'We believe that the game contains genetic material from blackjack. But we can't be sure. We didn't create the game, after all.'

2 **A crossword set by Dr Fill** Matthew Ginsberg has created an AI called Dr Fill that is better at *New York Times* crosswords than all but the top human solvers. It also sets clues itself and you can test your mettle below. (Answers at the end of this section.)

ACROSS

1. Most celebrated
6. 20's suppliers
10. Element in Einstein's formula
14. Noted clergyman
15. Unit of loudness
16. Graphic beginning
17. Peaks
18. Prefix with market
19. Sigmund's sword
20. It's legal in Massachusetts
23. Timorous
24. Data measure
25. Tend
27. Native-borne Israelis
32. The skinny
35. Type of skirt
37. Nonsense, slangily
38. Not-so-great explanation
41. Just around yon corner
42. "Groenlandia", e.g.
43. __ to the city
44. Slays
46. Wants
48. Oz dog
50. Kind of ax or ship
54. Video game featuring Gloom-shrooms, Melon-pults and Cherry Bombs
59. Grade
60. Cruising
61. Exuviates
62. Confess
63. Pastures
64. Where Rushdie's roots are
65. Vitamin A sources
66. Famous last words
67. Itsy-bitsy

DOWN

1. Women with __
2. 70's Renault
3. Ashlee Simpson album with the song "Boyfriend"
4. Convertibles that extend
5. Red Sox Nation's anthem
6. Culmination
7. Display contempt for
8. Brightly colored eel
9. Two jiggers
10. Men who made a star trek
11. Subtle quality
12. Prenuptials party
13. Not all
21. Anasarca
22. Extend to
26. Kon-___, Heyerdahl's boat
28. Sternum
29. Chess castle
30. Middle East port
31. Vodka sold in blue bottles
32. "If __ My Way," 1913 classic
33. One-billionth: Comb. form
34. Scamper away
36. Unemployed
39. Actress Ekland
40. Cardio option
45. Key on a cash register
47. "Nerts!"
49. Show case?
51. Diacritical mark
52. Admit
53. Test type
54. Say the rosary
55. Pumice
56. Energy source
57. Fresh reports
58. Silents star Pitts

3 **A video game created by Angelina** Space Station Invaders is a game in which you control a scientist who must fend off rogue robots and invading aliens to escape a space station. The artwork is by Angelina's creator and collaborator Michael Cook. But the layout of each of the levels, the enemy behaviour and the power-ups that give a player extra abilities were invented by Angelina.

You can play the game in your browser here:

https://www.newscientist.com/article/space-station-invaders/

You can play more of Angelina's games here:

www.gamesbyangelina.org/games/

4 **A recipe created by Chef Watson for Thai turkey strudel**
Serves 6

INGREDIENTS

450 g turkey

Frozen pastry

Half a seeded, minced Thai chilli

1¼ tsp rice flour

Dash lemongrass

Green curry paste

1¾ head lettuce

500 g potato, chopped

13 spring onions, chopped

1½ tsp vegetable oil

Olive oil spray

115 g Gruyère, diced

100 g Provolone cheese

SUGGESTED STEPS

   i Cook lettuce in boiling water.

  ii Drain and squeeze dry.

 iii Heat vegetable oil.

 iv Add spring onions and Thai chili and sauté for about 7 minutes.

  v Finely chop turkey, cheeses, lemongrass and rice flour.

 vi Transfer to bowl and stir in spring onions, lettuce and potato.

 vii Season with salt and pepper.

viii Preheat oven to 180 °C.

 ix Spray large baking sheet with oil.

  x Stack pastry in layers and spray with olive oil.

 xi Spread turkey mixture down centre of pastry.

 xii Fold short sides of pastry over filling, then roll up into log.

xiii Bake for about 40 minutes.

xiv Spoon green curry paste on the side and serve.

Eleven iconic AI villains

1 **False Maria,** *Metropolis* **(1927)** One of the first robots ever depicted in film, False Maria is a *Machinenmensch* or 'machine person' built by a brilliant scientist to impersonate a woman called Maria. But False Maria ends up bringing down the city of Metropolis by inciting its citizens to kill each other and destroy the city's machines.

2 **HAL 9000,** *2001: A Space Odyssey* **(1968)** The Heuristically Programmed Algorithmic Computer, aka HAL, is the AI on board the spaceship *Discovery One*. Unable to resolve conflicting mission objectives, HAL vents the ship and kills most of the crew before it can be shut down.

3 **Ash,** *Alien* **(1979)** Ash is the science officer on board the ship *Nostromo*. He appears human, revealing himself to be an android only late in the film. His secret mission is to bring the alien life form back to Earth.

4 **Roy Batty,** *Bladerunner* **(1982)** Batty is a replicant – a human-like android like Ash – who wants to extend his lifespan. Told that this is impossible, Batty kills his maker.

5 **Skynet,** *Terminator* **(1984)** The mastermind behind the machines in the *Terminator* films, Skynet is an AI system that becomes sentient after spreading itself across computers around the world. Civilization-ending war inevitably follows.

6 **ED209,** *RoboCop* **(1987)** The Enforcement Droid Series 209 – or 'Ed 209' – is a heavily armed police

robot designed to 'disarm and arrest' criminals. Its low intelligence and frequent malfunctions mean most encounters end badly for human targets.

7 **SHODAN,** *System Shock* **(1994)** An AI with a god complex, the Sentient Hyper-Optimized Data Access Network, aka SHODAN, controls the space station Citadel. After a hacker deletes SHODAN's ethical constraints, the AI becomes a megalomaniac and the chief antagonist in this horror video game.

8 **The Machines,** *The Matrix* **(1999)** The machines have plugged every human into the Matrix where they live in a near-perfect simulation of the real world as it was – while their bodies are harvested for heat and energy.

9 **The Cylons,** *Battlestar Galactica* **(1978–9; 2004–9)** Originally clanking metallic robots, the new breed of Cylons are indistinguishable from humans. Their determination to chase the last of humanity across the galaxy to wipe them out is the same, however.

10 **GLaDOS,** *Portal* **(2007)** An AI that guides the player through the strange test lab in the *Portal* video games, the Genetic Lifeform and Disk Operating System, aka GLaDOS, slowly reveals its true colours – and intent to kill the player.

11 **Maeve Millay,** *Westworld* **(2016–)** At first, none of the human-like robots in the theme park Westworld know they're machines. But after years of mistreatment for the sake of cruel entertainment for the rich, some start to gain awareness. Maeve Millay is the first to break out of the park's confines, killing the humans in her way – not that you can blame her …

Six computer-generated jokes

A team at the University of Aberdeen, UK, created the Joking Computer as part of an investigation into what makes jokes funny. Here are six of its best:

1 What do you get when you cross a frog with a road? A main toad.

2 What kind of a temperature is a son? A boy-ling point.

3 What kind of tree is nauseated? A sick-amore.

4 What do you call a cross between a bun and a character? A minor roll.

5 What do you call a shout with a window? A computer scream.

6 What do you call a washing machine with a September? An autumn-atic washer.

Six places to read more

1 Alan Turing's 1950 paper 'Computing machinery and intelligence' is where the field was born. In it, he considers the question 'Can machines think?' and lays out the rules for his Imitation Game. A pdf of the paper is available from many places online if you search for it.

2 OpenAI Blog, blog.openai.com

3 Google Research Blog, research.googleblog.com

4 Facebook Research Blog, research.fb.com/blog

5 Amazon Web Services AI Blog, aws.amazon.com /blogs/ai

6 Stuart Russell and Peter Norvig's *Artificial Intelligence: A Modern Approach* (Pearson, 2013)

Nine ways it could all end very badly

In 2016 Roman Yampolskiy, a computer scientist at the University of Louisville in Kentucky, and hacktivist and entrepreneur Federico Pistono published a list of worst-case scenarios for what a future malevolent AI might do. Here they are, in ascending order of terribleness:

1 Take over resources such as money, land and water.

2 Take over local and federal governments and international corporations.

3 Set up a total surveillance state, reducing any notion of privacy to zero, including privacy of thought.

4 Force merger (cyborgization) by requiring that all people have a brain implant that allows for direct mind control/override by the superintelligence.

5 Enslave humankind by restricting our freedom to move or otherwise choose what to do with our bodies and minds. This can be accomplished through forced cryonics or concentration camps.

6 Abuse and torture humankind, with perfect insight into our physiology to maximize the amount of physical or emotional pain, perhaps combining it with a simulated model of us to make the process infinitely long.

7 Commit specicide against humankind.

8 Destroy or irreversibly change the planet, a significant portion of the solar system, or even the universe.

9 Given that a superintelligence is capable of inventing dangers we are not capable of predicting, it could do something even worse that we are incapable of imagining.

Crossword answers

Across

1. A-list 6. ATMs 10. Mass 14. Peale. 15. Phon 16. Auto
17. Acmes 18. Euro 19. Gram 20. Same-sex marriage 23. Trepid
24. Byte 25. See to 27. Sabras 32. Info 35. Mini 37. Crock
38. Half-baked theory 41. Anear 42. Isla 43. A key 44. Does in
46. Needs 48. Toto 50. Battle 54. Plants vs Zombies 59. Rate
60. Asea 61. Molts 62. Avow 63. Leas 64. India 65. Yams 66. Et tu
67. Teeny

Down

1. A past 2. Le Car 3. I am me 4. Sleep sofas 5. Tessie 6. Apex
7. Thumb one's nose at 8. Moray 9. Snorts 10. Magi 11. Aura
12. Stag 13. Some 21. Edema 22. Reach 26. Tiki 28. Breastbone
29. Rook 30. Acre 31. Skyy 32. I had 33. Nano 34. Flee 36. Idle
39. Britt 40. Taebo 45. No sale 47. Dammit 49. TV set 51. Tilde
52. Let in 53. Essay 54. Pray 55. Lava 56. Atom 57. News 58. ZaSu

Glossary

Deep learning – a form of machine learning that uses neural networks with many layers

Evolutionary or **genetic algorithms** – software that tries to converge on an optimal solution by repeatedly combining the best work-in-progress solutions over many iterations, mimicking natural selection

General artificial intelligence – AI that can carry out a wide range of tasks with human-like ability

Narrow artificial intelligence – AI that is good at a specific task only, such as picking out faces in a crowd or driving a car

Neural networks – software circuits loosely based on the structure of the brain

Reinforcement learning – training a neural network by giving it positive or negative rewards for its actions

Supervised learning – training using data that has been labelled or annotated, e.g. a photo of a cat tagged with 'cat'

Unsupervised learning – training using data that has no label attached, e.g. a photo of a cat with nothing to say what it is

Picture credits

All images © *New Scientist* except for the following:

Figure 1.2: Sipa Press/REX/Shutterstock

Figure 2.1: Tom Zahavy, Nir Ben Zrihem and Shie Mannor

Figure 3.1: UNG YEON-JE/AFP/Getty Images

Figure 4.1: GATEway Project

Figure 4.2: Rockstar Games

Figure 4.3: Moviestore/REX/Shutterstock

Figure 5.2: Kupferman/CSM/REX/Shutterstock

Figure 6.1: www.thepaintingfool.com

Figure 6.2: www.thepaintingfool.com

Figure 6.3: www.thepaintingfool.com

Figure 7.1: China News/REX/Shutterstock

Figure 8.1: Adapted from Murray Shanahan, Imperial College London

Figure 8.2: Hulya Ozkok/REX/Shutterstock

Index

Interested in learning more?

Learn more about the world and the big issues affecting us by downloading New Scientist Instant Expert audiobooks and ebooks today.

All of the New Scientist Instant Expert audiobooks and ebooks in this series are available to purchase from the Instant Expert app and from instantexpert.johnmurraylearning.com

Use **NSIE40** at instantexpert.johnmurraylearning.com for 40% off any purchase.